cellular wisdom

cellular

wisdom

DECODING THE BODY'S SECRET LANGUAGE

Joan C. King, Ph.D.

CELESTIAL ARTS
Berkeley | Toronto

Celestial Arts
P.O. Box 7123
Berkeley, California 94707
www.tenspeed.com

Distributed in Australia by Simon and Schuster Australia, in Canada by Ten Speed Press Canada, in New Zealand by Southern Publishers Group, in South Africa by Real Books, and in the United Kingdom and Europe by Airlift Book Company.

Jacket design by Jeff Clark at Wilsted & Taylor Publishing Services
Text design by Nancy Austin
Jacket cover photograph of LHRH cells © 2004 by Joan C. King

Library of Congress Cataloging-in-Publication Data
King, Joan C.
  Cellular wisdom : decoding the body's secret language / Joan C.
King.
      p. cm.
    Includes bibliographical references and index.
    ISBN 1-58761-188-0
  1. Self-actualization (Psychology) 2. Physiology—Miscellanea.
I. Title.
  BF637.S4K548 2003
    158—dc22   2003016394

First printing, 2004
Printed in the United States of America

1 2 3 4 5 6 7 8 9 10 — 08 07 06 05 04

To Arnie Gerall, whose light burns brightly.

ACKNOWLEDGMENTS

The ideas expressed in *Cellular Wisdom* emerged from a new view of science growing within me, encouraged by the enthusiasm and insight of many. Brenda Rosen literally "birthed" *Cellular Wisdom* as a midwife, guiding the development of the book. Friends and colleagues participated in focus groups generously exploring how the ideas of *Cellular Wisdom* give insight into ways to lead exuberant lives. Thanks are due to those who took time to read the manuscript and provide feedback. Jean Houston's appreciation for ever-new emerging ideas and that of her co-leaders Peggy Rubin, Betty Rothenberger, and the participants of the 1998 and 1999 "Mystery Schools" provided opportunities to explore the principles underlying cellular wisdom. The Chicago Medical Society, Ruth Inge-Heinze's conference on the "Study of Shamanism and Alternate Modes of Healing," the Horizon Institute, the Science and Consciousness Conference and the Theological Opportunities Program at Harvard's Divinity School, each provided opportunities to develop cellular wisdom in distinct ways.

Many teachers of science, of spirit, of life, and of writing tilled the soil out of which cellular wisdom emerged. I am grateful to all of you for your scholarship, devotion to discovery, relentless pursuit of the unseen and passion for teaching others to see what you see.

I thank Celestial Arts publisher Jo Ann Deck for her support and editor Meghan Keeffe and copyeditor Heidi Garfield for *Cellular Wisdom*'s final transformation into a book for all people. As always, I thank my husband, Stuart Tobet, for his unwavering support of my creative explorations and his valuing of my expression of what I discover. Finally, I need to thank Mr. Nubbs and his partner Stache for making coming down the home stretch a continuously evolving novel experience.

CONTENTS

PART I:

living exuberantly within the membrane

what is cellular wisdom?

PART II:

living expansively in community

what is community?

INTRODUCTION

decoding the body's secret language

FOR MORE THAN TWENTY YEARS, I have studied the physical workings of the body as a research scientist and medical school professor. My specialty is neuroscience, the structure and function of the brain, spinal cord, and nerves. My personal research has focused on the dynamic interplay between the brain and the pituitary gland in the control of reproduction. Basic science—experimental studies, laboratory research, hard data and their analyses—have been the center of my professional life.

Yet in the past several years, an amazing thing has happened. My research lab at Tufts University School of Medicine in Boston became much more than a temple of science. I discovered it to be a temple of the human spirit as well. I began to see, in ways that went beyond everything that I had been taught and everything that I was teaching in my courses in medical neurosciences, that the physiological processes of the body—interactions within and between individual cells, organs, and system—had significance beyond human physiology, beyond the physical plane altogether. I began to see that fundamental human truths, a primer, even, for living an authentic and self-fulfilling life, were written in the cells, spelled out in the synapses, outlined in the interactions between the body's organs and organ systems. The basic science I knew opened like a flower, and I became more than the research scientist, professor, and administrator I had been.

Cellular Wisdom is a window into the truths I discovered, the lessons that our body can teach us about the basic principles on which the human spirit operates and moves and breathes. It shares the "aha!" understandings that emerged for me about how to live our lives, the principles that are implicit in the biological principles of human neuroscience. It reveals the body's cellular wisdom—the life lessons implicit in the secret language of the body.

My understanding of the body and its processes shifted in these past few years, taking on unfamiliar shapes—beginning, even, to shine and pulse with new meaning. Cellular wisdom continues to bring a new awareness of previously unseen promise, fresh ways of viewing the fundamentals of one's life, one's relationship with self, and relationships with others.

Stepping Off into Space

It was the summer of 1996. My husband and I had both given scientific talks at an international conference on hormones, brain, and behavior in Turin, Italy. I had delivered a paper detailing my research into the mechanisms governing neuronal interactions in the reproductive cycles of female rats. We were on the way to Rouen, France, where we would attend another international meeting and give another set of presentations. The neuroscience paper I was scheduled to present in a few days was research-based and abstruse.

For the moment, however, we were on holiday, touring the lush, green French countryside. We had stopped for the night at the Hotel Beau Site in Tailloire. At dinner that evening in the candlelit dining room, I overheard a young, English-speaking woman telling her husband how terrified she'd been watching him parasail that morning.

The word parasail reverberated in my consciousness. I'd always been fascinated by the sight of people floating beneath colorful, billowing sails, suspended high in the air. Something clicked inside of me, and I said to my husband, "There's parasailing here. Before we leave tomorrow morning, I want to try it."

My husband looked at me as though I had gone mad, and I guess in a way I had. Generally, I'm paralyzed by heights. Moreover, I can't swim. Neither of us could visualize my actually jumping off a cliff and sailing into the air over Lake Annecy. Yet something inside me told me I had to do it. I needed to perform a daring action, to demonstrate that I was ready to break free of my old way of being.

Why was I compelled to do this crazy thing? I was a respected neuroscientist with a long list of publications to her name. Chair of the Department of Anatomy and Cellular Biology and director of a research center at Tufts University School of Medicine, I had reached what many would call the pinnacle of success. But I felt overwhelmed by the demands of running a large university department, juggling three budgets, and responding to the exacting requests of three deans, while maintaining an active research lab. Often, I felt caught between what my head told me had to be done for the good of the department, and what my heart told me the faculty would feel about the changes I, as an administrator, was forced to make. More than once, returning home after an exhausting and contentious day, I spent the evening in tears.

Academia had been my home. It was more than a job; it was a way of life. Teaching, research, and giving talks at scientific meetings like the ones on this trip were all I knew. Consequently, when thoughts of leaving that life entered my head, I berated myself. Who would be foolish enough to leave a tenured professorship?

Yet everything I had learned about neuroscience was telling me that it was time to make a change. I had seen in the laboratory that the smallest building blocks of life, the cells which make up the organs and systems of the body, direct their activities from a centrally located nucleus. I had come to believe that this same principle—that living systems function by directing their activities from their core—applies to life at every level. I knew, without question, that if I did not begin to live from the center of my being, I would collapse, or explode from the tension.

And so, at five the next morning, we drove to the top of the mountain, Le Col de la Forclaz, overlooking Lake Annecy. My husband was humoring me, I guess, figuring I'd back out at the last minute. I scrunched down in my seat, barely daring to look out of the car window at the view below.

As I suited up at the site, I asked the man in charge, "Are people afraid to do this?"

"Lady," he replied, "anyone who hasn't done this before is afraid."

I took a deep breath, closed my eyes, and ran toward the cliff. Suddenly the air swooped under my sails, and I was lifted into the air. I had let go of the earth, yet invisible forces still supported and guided me from the moment I untethered and stepped off into space.

That day was September 1, 1996. Exactly one year later, I untethered again. I resigned my position as department chair and took a sabbatical, which I then extended by a year's leave. I was flying toward an unknown future in which science, in a way I could not yet see, would direct my activities from the core. Invisible forces under my wings would have to show me the way.

Biology as a Source of Wisdom

My first hint that the principles of hard science held the keys to human mysteries came in the early 1970s, when I was a graduate student at Tulane. I had been taught that neurons, the cells which make up the brain and spinal cord, were the most stable cells in the body. The neurons of an adult animal might die, but they did not change their structure. Yet as I peered through an electron microscope at the neurons of the brains of the adult female rats I was studying, I saw regular cyclic variations within individual cells. Organelles, miniature organs within the cell, seemed to appear and then disappear in a regular and repeatable pattern over a four-day period.

I tried to discuss these findings with my professors. The response was always skeptical. "Expand your sample," I was told. "Do more animals." Experimental research involves months of time: time to breed animals, time waiting for them to be born and to reach puberty, time to take vaginal smears over several months to determine when they established regular estrous cycles. Then, there is sampling animals on each day of the cycle, preparing the tissues, cutting brain sections thinly enough for an electron beam to penetrate, examining multiple sections in exactly the same region of the brain, taking photographs, developing the images, analyzing the data. All of this, and I could not know what the research would show until it was finished. I did more animals. The results were the same.

Convinced I was seeing something that other researchers had overlooked, I began to speculate about what might be causing the organelles to appear and disappear. I had injected one group of female rats in my study with the male hormone testosterone when they were five days old. As I expected, my tests showed that these rats had abnormal ovaries and were incapable of cycling. The mysterious organelle I was studying was apparent in their brain samples. I ran the same tests on a control group of rats whose estrous cycle was normal. Their brain samples also showed the organelle, but with a significant difference. The organelle appeared and disappeared at regular intervals. Not only were the rats' ovaries affected by early exposure to a hormone, but their brains were different, too. In normal adults animals, I was forced to conclude, the physical structure of the cells of the brain can change. The brain, as science describes it today is plastic.

The implications of what I was seeing were profound. If the brain is malleable, even this most stable aspect of our physiology can change in response to changing conditions. Under the microscope, I had seen neuronal changes in response to the wash of ovarian hormones triggered during the monthly reproductive cycle. But what other conditions might cause changes? Does the brain change physically, for example, when hormones are removed or altered during menopause? What about other life events? What changes might they trigger?

I was aware, of course, of the dynamic principles that guide interactions between other organs, such as the heart, liver, and kidney, and between organ systems such as the digestive and circulatory systems. Organs and organ systems retain their high levels of responsiveness by changing, by turning off and turning on in response to conditions. If an organ were to remain stuck in the on position all the time, its sensitivity would decrease, and soon it would lose its ability to respond. This is also true of neurons, which are on only when they are actually conducting an impulse. I did not imagine, however, that this principle might influence physical structures within neurons. To what extent, I asked myself, is life at every level malleable, plastic?

Patterns repeated on many levels in the body usually indicate the presence of a basic underlying mechanism. Key principles are obvious in many different circumstances, at many different levels of the body's functioning. Take, for example, the principle that information flows from the interior of a cell to its outer reaches. The same pattern is repeated in the way instructions pass from the interior of a population of cells to cells on the periphery. Moreover, the principle underlying this pattern—that guidance originates from the center—can be seen everywhere in the body. Could it be, I wondered, that this pattern holds true in the larger mechanisms of personal and social interaction as well? Could it be a principle that we might follow to make our lives happier and more fulfilling?

I began to ask questions that pointed beyond my research in the lab, toward the principles that direct these larger life processes. How might we, like the physical components of our bodies, learn to change as a result of changing conditions so that we maintain our sensitivity and our ability to respond? Could the physiological truths of biology, the way cells and organs and organ systems operate, hold important lessons for how we live our lives, how we think and feel and make decisions? Most important, could biology be a source of wisdom that we might tap into for guidance?

Change from the Center

Looking back, it's clear that my life up to this time had prepared me for these speculations. Experience had taught me that the ability to respond from one's center to changing conditions is essential to optimal functioning. Immediately after high school, I had entered a convent of Dominican sisters in New Orleans. When I joined the order, I told the sister interviewing me that I loved science and had wanted, since childhood, to find out what makes people tick. The order did not deem psychology, my first choice of college majors, appropriate. Instead, the sisterhood sent me to study chemistry, biology, and earth sciences at St. Mary's Dominican College in New Orleans. Later, I went to Florida State University to study nuclear chemistry and physics. As soon as I completed my studies, I began teaching at St. Mary's.

The pace of my academic and religious life was unforgiving. Up every morning at five for chanting and meditation, then off to my eight o'clock class. By five I was back at the convent for more chanting, dinner, and evening prayers. During the after-dinner recreation period, I was often too exhausted to speak. Long before lights out at ten, I was asleep. Summers were equally hectic, as I was sent off to graduate courses at various institutes and universities. Clearly, the system of my life was stuck in the on position. I was in danger of losing my sensitivity and my ability to respond.

Though some days I felt like I was running a race I couldn't win, it never occurred to me that I could leave the convent. Then one night, during an animated conversation in which I expressed my unhappiness with the unbending rules of the convent, one of the sisters said to me, "Well, then, if that's the way you feel, why don't you leave?"

That night, after lights out, her words reverberated in my mind, gaining momentum through the night. Finally, it dawned on me. Change, fundamental change that would alter the very structure of my life, was possible—even necessary—if I wanted to stay vital and keep growing. By the next morning, I was dancing around my room with a new chant on my lips: "I can leave, I can leave, I can leave."

When I told the Mother Superior that I would be leaving at the end of the semester, she was astonished. I was an excellent teacher and had been an exemplary nun. Eleven years after entering the convent, I left. Within a year, I began to study psychology and neuroscience, earning a master's degree at the University of New Orleans and a Ph.D. in neuroscience from Tulane. After post-doctoral studies, I was hired as an assistant professor at Tufts University in Boston and began teaching neuroscience to medical students. I found that I enjoyed simplifying the principles of the brain's functioning so that students could readily understand them.

My lab research at Tufts also focused on uncovering the parameters underlying the brain's operations. I probed the principles that directed the activities of individual neurons and neurons linked in circuits. I selected a particular set of neurons to study. "My" neurons synthesized a hormone that controlled the pituitary gland. That hormone (LHRH) acted on the pituitary gland causing it to secrete another hormone (LH).

For months my research associates and I meticulously mapped and recorded the position of each neuron that made LHRH in the brains of the lab rats. We color-coded the neurons according to different conditions. Viewing them on the computer screen, cells that made LHRH during one condition would show up green, those that made the hormone during a second condition would be red, and those producing LHRH during a third condition would be blue.

After months of work, we entered the final keystrokes and began to view the results. On the screen before us, we saw a three-dimensional model of the LHRH-synthesizing neurons. Much to our surprise, the neurons in the center of the brain were white, even though this was not one of the colors we had coded. Surrounding the white cells were others that were distinctly red, green, or blue. Each of the colored cells showed up in one of the conditions but not in the others.

Then we got it! The central cells showed up as white because white is the combination of red, green, and blue. The white cells in the center of the brain seemed to be activated in all conditions. Could it be that the neurons in the center of the population were like a light switch, turning on other neurons as needed?

Body systems also have central control mechanisms; for example, the brain controls the nervous system, the heart, the circulatory system. The job of the central control center is to provide a seamless flow of information to coordinate activity within a cell, between cells, and between organs and systems.

Without such control centers, communication breaks down, vital functions do not take place, and disease or death results. Is this also what happens when we operate from fragmented pieces of ourselves rather than from our core? The elegant mechanisms of the body underscore that to be healthy and whole, we must live our lives from the core essence of who we are.

The Body's Secret Language

The research work I was doing in the lab simply wouldn't stay there. It kept spilling over into my life. I thought about a friend who had recently confided that she was leaving her husband of many years. Pushed by her father to excel in business, this woman had become a top financial planner. The man she married had a similar drive for success, and the couple built a life focused around their high-powered careers.

Then, suddenly, the structure of my friend's life began to crumble. No longer able to endure the pain of dragging herself to work each day, forcing her mind to focus on how to make her rich clients more and more money, she left her secure position in a respected investment firm. Wanting to help people, ordinary people, understand how to leverage their income, she started her own small consulting business. One change led to others, and within a year, she had decided to separate from her husband.

Why, I wondered, did both her first career and her personal life dry up at the same time? Could it be that she had been living her life from the periphery and not from the core? I could not help but compare my friend's story to what I had seen in the lab. Once I began to see the pattern, I noticed it everywhere—in my life, in the lives and experiences of my friends, and in the ways we all interacted with the institutions, corporations, and social organizations in our lives. I began to believe that the principles that operate in the body can be meaningful guides to constructing authentic and self-fulfilling lives.

I thought, for example, about the fact that many organizations become increasingly ineffective over time. I wondered if this diminishment was related to the loss of a core vision. Steve Jobs catapulted Apple computers into the mainstream of American schools and homes in the 1990s. Not too long after Jobs left the company, Apple Computers almost died. Jobs returned and despite dire predictions, Apple once again soared to prominence. As members of an organization, each with their own agenda, begin pulling the organization in directions that are not aligned with its vision, the energy of the organization's original focus dissipates, and effort fragments. Eventually, the organization breaks down. It is as if the animating spirit of the institution has dried up, and the institution can no longer thrive.

My own life corroborated my speculations. The multiple demands of my professional life as a teacher, researcher, and university administrator seemed to be fragmenting me, strangling my creativity, drying up my spirit from within. I spent my sabbatical and the year-long leave that followed thinking and writing about these issues. I wanted to get to know myself again. Who was I underneath my achievements, skills, and competencies? What did I really want out of my life? Where did I want my life to go? As a part of this process, I consulted with a personal and professional coach. She encouraged me to ask myself difficult questions and to formulate a plan for rekindling my passion for life.

After eighteen months of hard thinking, my path was clear. I had looked inside and consulted my core values. They pointed me toward starting a coaching business of my own, through which I might help others rediscover how to live exuberantly from the center of their being. I called my business Beyond Success to indicate that success was not the answer to the yearning of the spirit. As a coach, my specialty would be teaching people to apply the dynamic principles of biology and physiology to their personal struggles and to their interactions with others.

My training in the laboratory gave me many ideas about how to proceed. I organized a focus group to test the principles I was formulating. Time and again, my living room research bore out the validity of the methodology I was developing. I recall one evening when the topic of fear seem to dominate the conversation. I described to the group one way that animals handle fear.

"If you turn an animal quickly onto its back, it freezes." I said. "The condition is called tonic immobility."

One of the group members, a woman named Naomi, sighed audibly. "The same thing happens to me!" she said. "I've been tearing myself up over it for years. When I'm afraid, I just freeze. I had no idea my reaction had a name. I just thought I was a person who couldn't be counted on to think clearly when something scary is going on."

Heads nodded throughout the room. The sense of relief in the group was palpable. Knowing that tonic immobility is a physiological response to having one's world turned upside down gave them permission to relax. We talked about how freezing up creates a natural pause, allowing our competing responses to calm down so that we can tune in, think more clearly, and marshal our forces for action. The conversation became animated as person after person recounted situations in which they could have benefited from this knowledge. How reassuring it was to learn that we could follow the lead of the body in handling fearful situations.

Your Body's Truths

Books about mind-body interactions are popular today. We have come to understand that physical illnesses often have messages for us. In fact, our thoughts and emotions affect our health and the way our bodies function.

The concept of cellular wisdom takes the mind-body connection a step further. It invites you to consider that the principles of physiology—the basic biology of the body—contain key truths that you can use to create an authentic, fulfilling life. Cellular wisdom uses these biological principles as a basis for addressing the following question: How can you make all the components of your life—your body, emotions, psyche, and spirit—function as a harmonious and synchronous whole?

The first section of this book looks at the principles that operate within the membranes of a person's individual body, within and between the cells, organs, and organ systems that keep each of us healthy and functioning with ease and joy. The second section of the book examines the principles that guide our lives as social beings—one-on-one relationships with a partner, family member, friend, or colleague; interactions with the organizations, companies, and other groups in which people have memberships; and the relationships with the institutions and other systems that impact our lives in community. Cellular wisdom, you will discover, is wisdom written in the body. Lessons for exuberant living are coded into the elegantly complex dance of chemical and energetic interactions—the physiological component of your every thought, word, and deed.

You are the scientist of your life. As you explore problems and decisions from the perspectives offered here, you may see alternatives that had not suggested themselves to you previously. If you choose to test these new ways of thinking and acting in your life, you can rest secure in the knowledge that the principles that guide you come from a most reputable source—the truths that your own body teaches.

PART I

living exuberantly within the membrane

WHAT IS CELLULAR WISDOM?

Every cell in your body is a genius!

Encoded into its molecules is the ancient wisdom of the earliest cells, refined over billions of years, the energetic essence of our evolutionary heritage. Eons ago, before the first cells evolved, molecules synthesized in a random fashion. There is no memory of these events. They simply occurred from time to time, with no order, no permanency. Over time, perhaps a billion years, simple molecules evolved into complex ones, until, about four billion years ago, the first cells took form.

When the complex molecules of the nucleic acids, DNA and RNA, came into being, memory was created, as these molecules had the gift of self-replication. Mitochondria entered the cells to function like power plants, turning molecular oxygen into fuel that cells use.

In the next movement of the evolutionary dance, cells joined together to form simple creatures composed of a single cell type, such as sponges. Step by step, such one-dimensional beings evolved into complex life forms, composed of many different types of cells—brain cells, bone cells, muscle cells. Groups of similar cells joined together to form cellular communities, and tissues, organs, and organ systems came into being.

We humans are the inheritors of this evolutionary genius. The creative energy from which the first cell emerged has not been lost. It hums within the coding of every cell of our body.

The life of each cell, whatever its function, is orchestrated from within by an abundance of stored information. Every moment of every day, the evolutionary blueprint coded into DNA and RNA is translated effectively and harmoniously into a cell's sustaining and communicating activities. The synchronicity of molecular events within cells and between cells is a symphony of reliable and repeatable events.

The tens of thousands of genes archived in the cell nucleus are not all turned on simultaneously. Rather, they are called to play their parts at specific times, at the precise moment they are needed. The marvelous dance of life in which our cells are engaged at every moment is the physiological substrate of cellular wisdom.

Yet the music that infuses cellular wisdom goes beyond physiology. Physics teaches us that, at its essence, everything is energy. We humans are beings of energy. If you've ever been present at the moment of death, you've witnessed the potency of this continuum of energy. A mysterious life force, made up of the source energy that reverberates within each molecule in every cell in our body, withdraws as we die. When this life force dissipates, energy no longer courses through our neurons, generating thoughts and motivating interactions.

Source energy that animates every living being was available before the emergence of subatomic particles, molecules, or cells. Eternally creative, ever seeking new dimensions, it continues as energetic force relentlessly pursuing new expressions.

Like all living beings, cells cannot thrive in isolation. Although they have boundaries—membranes that define their limits—much of each cell's activity is directed to communicating with other cells. Billions of years ago, as cells differentiated to generate a variety of cell types, they learned to signal each other by releasing specifically designed molecules. Once communication was established between cells, the activities of cells could be coordinated. Organs and organ systems evolved when cells learned to act cooperatively and harmoniously.

Our ability to communicate has given us the vast repertoire of behaviors characteristic of complex beings. Most profoundly, the source energy that resonates as communication between cells, and between beings composed of communicating cells, makes possible the miracle of interactive transformation. The miracle lies in the synchrony—one cell type, such as muscle or nerve, each at a certain stage of development, provides the cues to initiate the transformation of the other. Without this interaction, transformation would not occur. The energy waves that allow for communication and interconnection make up the expansive dimension of cellular wisdom.

When we open ourselves to source energy, we connect with cellular wisdom, for source energy is our link to the creative energies of the universe. We humans are an aggregate of vibrantly alive cells. We thrive, as have cells for billions of years, when our actions and interactions are aligned with the energy that animates and sustains our cells.

What does an experience of cellular wisdom feel like? Its most vivid and dramatic manifestation occurs in times of transition and change. Think back to a time in your life when the existing order broke down, and an unknown future was gestating. Maybe it was a time when you changed careers, separated from a partner, lost a child or a parent. Suddenly, life was different, marked by a different texture, different emotions, a different way of thinking. A special energy, which we recognize as part of our deepest self, characterizes these times. It is the energy of the void, the energy that powers all creation, the energy that resonates within our cells at every moment. It may make us light-headed or giddy. We may laugh or cry more easily. We may feel a hollowness at the pit of our stomach, a clutching sensation in our chest. These are physiological expressions of the void that marks the impending emergence of a new dimension of life. In the energy vortex of the void, we may feel lost. In a way, we are lost—to an old way of being.

Yet out of this void, something new is being born. A new reality is unfolding from the field of possibility, perhaps accompanied by a sense of loss; perhaps, by a sense of wonder. At moments of transition, the vortex of creative energy that is our life force is activated. Its waves carry the blueprint that informs every new aspect of our life. If we surrender to the vortex of energy, cellular wisdom will direct us on a path that leads to emerging new dimensions of our lives, for the energy of all new creation is cellular wisdom in action.

But dramatic transitions are not the only time we can experience the source energies of cellular wisdom. This energy is available to us whenever we are fully conscious of the present moment, because the present moment is the only place this energy exists. We cannot experience source energy in past events or in reveries about how we felt at some former time. We cannot connect with it in future events or in fantasies about how we may be months or years hence. Moreover, we cannot find this energy in our minds, because our thoughts are not our actual experience. Nor can we feel it when our emotions are agitated. Only in a peaceful experience of the present moment can we enter the stillness and connect with the energetic hum of our being.

There are many things we can do to relax into the now and connect with the energy at our core. Some of us access this state while engaging in exercise; others by listening to music, painting, or viewing a magnificent sunset. Meditative and contemplative spiritual traditions teach many techniques for quieting thoughts and tuning into ourselves. We know when we've made the connection, for we experience the sense that all is well.

We can drop into this state instantly, suspended between tasks, even in the midst of a busy day. No one need know that we are experiencing the energy that pulses through our bodies at the deepest level. Frequent connection with source energy gives us access to that state of wholeness and joy which is the harmony underlying all activities. As you will discover in what follows, connection to source energy is not difficult, because cellular wisdom is always present, always accessible. Quite simply, we tune into it by becoming aware of its presence.

Moreover, we elicit the connection to cellular wisdom by yearning for it. Yearning focuses and aligns our energies. Many of us think of ourselves based on images that we developed years ago, images that no longer apply. When we love who we are now, with all our imperfections, we actively open to source energy within us. The energy that animates us at the cellular level unfolds our spirit, moment by moment. Since we last spent time with ourselves, we may have added new dimensions. What draws us today may not be what drew us yesterday.

Any journey within leads, paradoxically, to heightened awareness of the expansive energy which permeates all life. The energy we encounter when we look inside ourselves is the same energy that guides the movement of the heavenly spheres—the stars, galaxies, universes. Each living being has a characteristic profile of energy frequencies expressed within his or her body, but our spirits are not confined by physical limits. Our individual energies pulses in and out of the sea of all energy, all love. When we tap into that confluence of energy, we arise renewed, knowing that the energies in our bodies are moving within the harmonies of vast cosmic music.

We each have an interior teacher who can help direct our lives, in the interior domain and with others, exteriorly. The goal of this book is to introduce you to this teacher and help you to understand and recognize its messages. There is nothing that you have to do to earn the presence of cellular wisdom, for it is nothing less than the essence of who you are.

CHAPTER 1

live from the inside out

EACH OF THE BODY'S ONE HUNDRED TRILLION CELLS is a dynamic, living entity, continuously engaged in the energetic dance of life. The life of a cell is orchestrated from a central command center called the nucleus, located at the heart of the cell. Every moment of every day, the genetic information coded into the complex molecules of DNA stored in the nucleus directs the cell's activities.

The principle that information flows from the center of the cell to its outer reaches is true at every stage of our development and for every type of cell. The DNA in the nucleus of the simple, cellular being from which we each developed contains the code used to translate our distinct genetic information into molecules. Early in our development, cells differentiate into various types—muscle cells, bone cells, glandular cells, neurons, and so on. Each cell carries in its nucleus a copy of the original DNA blueprint.

When internal or external signals reach the cell, particular segments of the DNA code are activated, causing the cell to generate the specific molecules it needs to carry out its activities. Cellular activity is precise and appropriate: cells generate only those molecules called for by the activated segment of DNA and only those molecules that are appropriate to its type.

Simply put, from the moment of conception until we die, cells live from the inside out, responding from their core precisely and appropriately to the signals they receive.

Like cells, each of us has a blueprint at the heart of our being. As we become more aware of the dynamics of our lives—what works for us and what does not—our individual blueprint of truth and values becomes more and more clear to us. When we make decisions in accordance with this blueprint, we feel vital and fulfilled. When we make decisions that contradict this truth, or operate from fragmented pieces of ourselves rather than from our core, we feel frustrated and distressed.

As is true of cells, there are many different types of people, each with distinct strengths and abilities. We become more responsive to those internal and external signals that activate aspects of our inner blueprint as we continue our process of self-discovery. When we respond to signals that resonate with our core values, events unfold that evoke our essential selves, and we discover new opportunities for self-expression. Like cells, we thrive when we live energetically from the inside out.

Examine the Blueprint

The cell is to be our wisdom textbook, so let's begin with a simple tour of its structure and functioning. Each cell is a complex assembly of tens of thousands of molecules and larger structures, suspended in a gel-like fluid. The cell's outer boundary is a plasma membrane, the interface between the cell and its external environment. We might think of this membrane as the cell's skin and sense organs.

Inside, the cell is divided into several compartments. The membranes that separate these compartments specialize in transmitting signals between the cell's various "rooms." The nucleus, which houses DNA—the director of cellular activity—is also surrounded by a membrane that separates the command center from the gel-like fluid.

The mechanism through which the DNA blueprint directs the synthesis of needed molecules is quite complex. In simple terms, however, strands of DNA receive a signal telling them to open so that the blueprint they encode is accessible. Specialized enzymes in the nucleus read the code on the DNA strands and transcribe the code to another molecule, a type of RNA. The RNA then crosses the nuclear membrane to carry instructions from the DNA into the gel-like fluid.

In the fluid, other types of RNA translate the code and direct the synthesis of proteins with the help of other specialized cellular factory sites, or organelles. The molecules in these factories are each created in the right amount, at the right time, to serve a specific purpose. They are then sent to sites within or outside the cell. The energy to fuel these cellular processes is generated by yet another molecule, which is manufactured in another cellular factory, called the mitochondria. Other organelles within the cell carry out the task of breaking down molecules that are no longer needed.

Though greatly simplified, this brief tour illustrates the ceaseless dynamic balance that goes on within the cell as it responds to internal and external signals, synthesizing new materials needed for life, sending them to places where they perform their functions, and breaking down those materials that the cell no longer needs.

]00000[

LIKE OUR TRILLIONS OF CELLS, we are also a ceaseless hub of creative, communicative, and destructive activity. We take in signals from our own bodies and from the outside world, respond to them in creative ways, and discard those structures—ideas, feelings, sensations—that we no longer need. All too often in the rapid pace of modern life, however, we seem to bypass our central control mechanism, moving instantly from stimulus to response without consulting our inner blueprint. When we make decisions based on the expectations of a parent, spouse, or peer group, or act in a certain way because we have always done it that way, we are ignoring the first principle of cellular wisdom: *examine your blueprint and act accordingly.*

How can we become aware of the life instructions—the core values and beliefs that are coded into our essential blueprint? Any process that encourages introspection, including meditation, journaling, and creative or artistic expression, can help us get in touch with the essence of who we are. When we're caught up in the flow of life, an uneasy feeling or uncomfortable sense that something is out of synch can signal that we are not living life from the inside out.

I think, in this regard, of Cassandra, acknowledged by her peers as a most successful criminal lawyer. Now forty-four and divorced, Cassandra was becoming increasingly uncomfortable with the arguments she was making to defend her clients, especially those she sensed were guilty. Though she was finding it harder and harder to feel good about her work, she continued, as she had for many years, to craft winning arguments and chase down supporting witnesses.

As she interviewed the defendant in her latest case—a high school teacher who was accused of having sexual contact with a fifteen-year-old female student—Cassandra became deeply agitated. On the surface, the man seemed sincere in his outrage over the student's accusations. His principal called him an exemplary teacher, and he was well-liked by his colleagues. Yet, as she pushed to find more character witnesses and ask more questions, Cassandra sensed that something was not quite right. The man seemed too eager to concur with her suggestion that the young girl in the case was disturbed and given to fantasy. Slowly, as she had been trained to do, Cassandra developed a strategy to discredit the information that the girl provided. So convincing was her presentation in court that the teacher was acquitted of all charges. Cassandra was celebrated as a winner.

But Cassandra did not feel like a winner. In fact, she had never felt this distressed, even after her divorce. Knowing that every accused person deserves the best possible defense was no longer enough to justify her performance in the courtroom. When she passed the young girl's parents in the corridor after the verdict, her eyes filled with tears, and she could not meet their gaze. Cassandra's profession was no longer in synch with her values. She was not living according to her inner blueprint.

What might Cassandra do to get in touch with her core values and use them to guide her actions? In the course of this book, I'll suggest many strategies for putting cellular wisdom to practical use. Here's a process you can use to begin the process of examining your inner blueprint.

REVIEWING YOUR LIFE CHOICES

Find a quiet place where you will not be disturbed for an hour or more. Sit comfortably and take a few deep breaths. As you exhale, sigh audibly or say, "Aaaaaahhhhhh." Relax. Think about the answers to the following situations. If you wish, record your answers in a journal or notebook.

[] Think of an important choice or decision you made in the past year. Mentally compare how you felt before making the choice.

[] What emotions or thoughts did you have while making the choice?

[] How did you feel after making the choice?

[] Looking back on these events, how do you feel now about the choice you made?

Compare this recent choice to two or three previous life choices—
perhaps the choice of a college major or profession, the choice to begin
or end a relationship, the choice to buy a house or move to a different
city, the choice of a spiritual affiliation or an important set of social or
cultural activities.

[] What do you notice about the circumstances surrounding these
choices?

[] Which choices made you feel particularly happy and fulfilled?

[] What was it about these choices that made you feel this way?

[] Which, if any, made you feel uncomfortable?

[] What was it about the choices that led to the uncomfortable feeling?

[] What do you discern about the values or beliefs that led to the life
choices you've made?

[] What patterns do you notice?

[] If you could go back in time to when you made one of these choices,
what would you do differently? Why?

We come to know ourselves by discovering which choices lead to a sense
of deep fulfillment and joy and which make us feel uncomfortable and
unfulfilled. Those that make us happy generally resonate with our essential
blueprint. Those that make us uncomfortable and regretful may not.

Taking a moment to examine the choices you've made and how you feel
about them can help you begin the process of examining your inner blueprint and discovering the truth of who you are. In my own life, I have discovered how important it is to make choices based on core values. For
instance, when I was asked recently to serve as president of Women in Neuroscience (WIN), I took the time to investigate whether this position would
be compatible with my values.

WIN's stated mission is to promote the careers of women who work in a
challenging scientific field. Because my work as a coach and teacher focuses
on personal and career development for professionals, the fit seemed to be
a good one. But I also spent the time to assess how I was feeling about
taking on this added responsibility. I asked myself honestly whether the
position might feel like too much, given the other activities I value. Would
I come to resent the time I would have to spend attending meetings and

doing administrative chores, rather than seeing clients or working on my writing? No red flags popped up.

In spite of the prospect of my being away from home more often, my husband gave me his full support. I then had a series of conversations with the woman who was then president of WIN. She reassured me that my new responsibilities would mesh easily with my current life, and that WIN's organizational culture would allow me to run the organization using a team approach, an important value for me. I agreed to take the job.

One of my first actions as president was to invite WIN's other officers to participate in a exercise designed to help us identify our personal values. After completing this exercise, we worked together to articulate a set of shared values that would guide us as a leadership team. Because I took the time to do my inner and outer homework, I have found my work with WIN to be satisfying and joyous.

You can probably think of instances in your own life when decisions you made had a similarly happy outcome. But you can probably think of other times when things did not turn out as well as you had hoped. In both cases, ask yourself how conscious you were of your process of decision making. If you discover that your habit is to make decisions by the seat of your pants or because one choice seems easier than another one, you've been risking disappointment. Make your core values—that inner blueprint so like the DNA at the heart of the cell—the principle that consciously guides your decision making. In doing so, you will better the odds that your choices will lead to positive and life-affirming results.

The Power of the Right Fit

Cells differ from each other in significant ways, structurally and functionally, yet each plays a vital role in the body. Bone cells, muscle cells, and cells in our connective tissue support the body's physical structure. Skin cells and cells that make up the sense organs help us interact with the external world. Cells in the pancreas make insulin; cells in the ovaries make estrogen; white blood cells in the lymph nodes make antibodies to fight infection. In each case, the cell's structure, vital functions, and life cycle are directed from within by the DNA blueprint in its central nucleus.

Let's look briefly at the lives of two very different types of cells—neurons and red blood cells.

The job of neurons is to transmit information. These cells develop before we are born and generally do not reproduce in adults. Although some are lost along the way, neurons are our constant companions from birth to death. Neurons are linked into intricate networks. Messages passing between them are carried by an array of neurotransmitters, which are specialized molecules that cross cell boundaries to carry impulses to other cells. Each neuron is a vital link in the transfer and integration of information.

Say, for example, that you want to raise the index finger of your right hand. When the decision to initiate movement is made in the motor cortex of the brain, a motor neuron (we'll call it neuron 1) fires. An electrical signal travels along the membrane of neuron 1's long tail descending through the spinal cord. In a person who is six feet tall, this single neuron can measure several feet in length. In a giraffe, by comparison, a similar motor neuron can be as long as six feet!

When the signal reaches the spinal cord, it is transformed from an electrical signal into a chemical one. A neurotransmitter is released from the ending of neuron 1 to transmit the chemical signal to a motor neuron (we'll call it neuron 2) in the spinal cord at the level of your shoulder.

Neuron 2 receives the input from neuron 1, along with input from as many as ten thousand other neurons that relay information about the tension in muscle and the state of the body. Neuron 2 takes in all this information, integrates it, and sends an electrical signal through its long tail down the right arm. When the signal reaches the index finger, neuron 2 transforms its electrical signal into a chemical one and releases its own neurotransmitter. This neurotransmitter activates receptors that cause contraction of the flexor and extensor muscles in the finger. The muscles contract, and your index finger moves.

In contrast to neurons, red blood cells are the simplest cells of the body. They have no nucleus or internal organelles. Essentially, they are disc-shaped balloons filled with hemoglobin and enzymes. Their simple design helps them perform one vital function with utmost efficiency: delivering oxygen and removing carbon dioxide from other cells in the body. And unlike neurons that are with us from before birth until death, red blood cells are short-lived. After delivering oxygen and removing carbon dioxide from other cells for three or four months, red blood cells are destroyed in the spleen by other cells that surround and digest them. In each second of life, five to ten million red blood cells are produced and destroyed!

Neurons and red blood cells represent the extremes of cellular life. Stable and nonreplicating neurons stay with us throughout life, while red blood cells are continually created in the bone marrow and destroyed in the spleen. The life cycle of each type of cell is particularly well suited to its function. If neurons were constantly being replaced, neural networks, such as the one described above between the brain and the muscles of the index finger, would continually have to be reconnected, which would disrupt the flow of information through the body.

]OOO[

LIKE CELLS, PEOPLE EXPRESS THEMSELVES IN VARIOUS WAYS. Some people thrive in interaction with others. Others require frequent retreats to solitude. Some people express themselves best with words; others use paint, or clay, or a musical instrument. There are those who come alive when they are gardening; others, when they are interacting with animals; and still others, when they are tinkering with machinery. Our inner blueprints encode the ways each of us most naturally thrives. When we spend most of our time engaging in activities that do not nurture us in essential ways—for instance, working at a job because it pays well rather than because it interests or excites us, or staying in a relationship because being alone feels too scary—we're like square pegs trying to fit ourselves into round holes. No wonder we feel out of sorts!

You probably already know a great deal about what you most enjoy and feel comfortable doing. But perhaps you've never taken the time to think about what these activities reveal about your coding and how you might harness the energy of your natural affinities to live more authentically from your core.

Cassandra, the defense attorney we met earlier, had always loved horses. Her favorite getaway during law school and when the stress of her busy practice wore her down was to retreat to her family's farm on the New England coast. Though she might feel withdrawn and edgy when she arrived at the farm, her mood would begin to shift as soon as she changed into riding clothes. Saddling up Maggie, the chestnut mare that had been her horse for many years, Cassandra headed for the fields. After riding hard for hours, she'd slowly begin to relax.

When she was tired of riding, Cassandra would lead Maggie back to the stable, remove her saddle, and brush her coat. While she was grooming the horse, Cassandra's breathing deepened, and any remaining tension she carried in her body dissipated. No matter what was going on in her professional or personal life, Cassandra always felt whole around Maggie.

On this day, soon after the victory she had won for the teacher she had defended, no amount of riding and grooming Maggie could make her troubling thoughts go away. From somewhere deep inside Cassandra, a voice was saying, "I can't practice law anymore. It's tearing me apart."

She recalled the many conversations she'd had with the father she adored, who'd been the one to urge her to take up law. Her father was delighted when Cassandra received her law degree and joined the firm in Boston. Whenever she was contesting a case, her father was her most enthusiastic cheerleader. Had Cassandra built a life on her father's belief that she would be an excellent lawyer? Was she living up to her own truth, or her father's expectations? She certainly was a terrific lawyer, she reminded herself, just as her father had predicted. Didn't that prove that her father was right? But, if he was right, why did she feel such turmoil, such unhappiness? And, if she didn't practice law, what would she do instead?

Cassandra spent many months considering these questions. Later that year, she took a leave of absence from her firm and traveled to Spain to spend time on a ranch, riding and thinking. In the end, Cassandra found that the only question worth asking was, What contribution do *I* want to make to the world? Though the answer did not come easily, an idea began to surface, a distant call like the sound Maggie made when she was contentedly munching oats while Cassandra groomed her.

Wouldn't it be wonderful if I could find some way to make a living breeding and raising horses? Cassandra mused. Then I could always feel the contentment I feel here on the ranch.

Activate Your Dreams

We have seen that the DNA in the nucleus of the cell controls the synthesis of neurotransmitters and other needed molecules within the cell. But what is the trigger that activates the DNA molecule and sets it in motion?

To understand the mechanisms by which the DNA blueprint can be activated, we have to look briefly at the structure of DNA itself. DNA is a large molecule that is folded and compacted along with other supporting molecules within a single chromosome. Chromosomes are rod-shaped structures apparent only in cells that are in the process of dividing. They carry the genes that convey hereditary characteristics and control other activities. In a cell that is not dividing, DNA is shaped like of a pair of twisted spirals, rather like the two sides of a ladder. Each spiral consists of a chain of complex molecules. The rungs of the ladder are not solid; only a portion of each rung is attached to one side of the ladder or the other. One strand carries the code for generating proteins, such as the enzymes required for neurons to synthesize neurotransmitters, or red blood cells required to synthesize hemoglobin. The other strand carries sequences that are used to replicate the DNA in cell division.

When enzymes or other proteins are needed, the strings of molecules that form the two sides of the ladder must separate so that the coded instructions can be read. Specialized enzymes then copy the sequences of molecules on the rungs and transcribe this code to a strand of RNA. The RNA carries the message of the DNA from the nucleus to the gel-like fluid in the cell, where it directs the synthesis of proteins. Of the multitude of coded instructions attached to a strand of DNA, only the part needed to accomplish a particular task is activated.

Let's look at an example. Many people have heard of melatonin. This chemical is produced naturally in the pineal gland, located between the two hemispheres of the brain. Melatonin helps the body regulate its day-night rhythms, and it is synthesized primarily during the night. The DNA sequence in the nucleus of pineal gland cells responsible for synthesizing melatonin is activated by special "clock" genes that turn on in a regular rhythm. Clock genes activate the DNA, which then directs the cell to synthesize the enzymes needed to make melatonin. Genes that turn on in response to light suppress the activity of the DNA during the day, thereby reducing the number of enzymes and, in turn, the amount of melatonin that the cell makes.

]0000[

LIKE THE DNA WITHIN OUR CELLS, we are each coded with many potentials that can be activated if conditions are right. These aspects of our inner blueprint lie dormant unless our expression, desire, and belief give them the opportunity to manifest. When we were young, we all dreamed possible futures for ourselves. Even as adults, most of us harbor a dream that lies sleeping at the heart of our being. Cellular wisdom can provide a model for what we must do to activate our dreams. If we imagine that our dreams are coiled deep within us, like the spiraling DNA blueprint at the heart of a cell, we must provide the trigger that opens the code and sets the process of manifestation in motion.

The trigger that activates our dreams is articulating them, first to our-selves and then to others. A dream that is shared—as the DNA in the nucleus shares its code with the RNA that transcribes it—can become a potent force for personal and communal change. We all remember, for example, the power unleashed by Martin Luther King's 1963 speech, in which the words "I have a dream . . ." ignited a nation. Because he spoke his dream out loud, a vast movement for transformation was born.

What is your biggest unspoken dream? Once you express your desire for a dream's fulfillment, your belief in your desire initiates a sequence of events that quite often takes on a life of its own. As if you have been turned on by a clock gene, your body moves into high gear. Desire and belief sensitize you to your environment, alerting you to ways to anchor your dream and translate it into reality. Synchronicities abound, and you recognize previ-ously unseen opportunities to take action. You begin to meet the right people, and they become allies or offer unexpected kinds of help. Soon a powerful process has begun that leads you to the synthesis of new ways of being or relating to others.

Remember the red blood cells we looked at in the previous section? These, too, teach us an important lesson about our dreams. The lives of these single-purpose cells are short because they do not carry within them a life-generating center—the nucleus with its DNA and RNA. When we for-get our dreams, we, too, are operating without a life-giving source of energy and motivation. Without dreams, our vitality eventually fades, and our lives lose direction and purpose.

It is never too late to waken a dormant dream and put it into motion. Martin, a scientist in a biomedical company, found himself doing less and less real science. Since being promoted to the post of administrator, team leader, and strategist, he found coming to work a chore, and he stopped looking forward to new projects. The one bright spot in Martin's work life was photography. When his administrative responsibilities gave him the opportunity to look at photographs of cells or to be present when the imag-ing of cells and other tissues was taking place, he was deeply engaged in a way that his managerial and strategic work never provided.

One evening, as he was reading the newspaper, he saw a photograph of a marathon winner that reminded him of a contest he had won with a sports photo he had taken for his college newspaper. He remembered loving the challenge of catching the faces of people, intense in activity, poised at the end of a springboard or coming up with a diving catch in the outfield.

Everyone commented that his images seemed to capture the powerful emotions of his subjects.

The next Saturday, he headed for the camera store to get some new equipment. He started carrying his camera with him wherever he went, seeking opportunities to photograph people doing ordinary things with extraordinary intensity. He even took pictures on the job and produced a photo spread, published in the company newsletter, showing scientists intent on a difficult experimental procedure. A friend who worked for the *Atlanta Journal-Constitution* admired the photograph Martin had taken of the face of a young woman who was a member of the research team. Martin's photo was featured in a *Journal-Constitution* article describing the team's experimental work on a new AIDS drug.

When his biotech company was bought and merged with another, Martin decided to take a risk and pursue photography as a profession. His photo in the *Journal-Constitution* had brought inquiries. Now, he was engaged in viewing the world through his camera as a freelance photographer. Whenever he was behind his camera, he felt a quiet joy. Taking pictures of people who were deeply engaged in their tasks came easily and offered him a kind of immersion in his work that his corporate responsibilities never provided. From a dream coiled deep within him that he dared to express, Martin manifested a new and fulfilling life for himself.

A Question of Timing

Martin was able to manifest his dream because a series of events occurred in sequence: the picture Martin saw in the newspaper, the photo spread he created for the company newsletter, the friend who placed Martin's picture in the Atlanta paper, and the buy-out of Martin's biomedical firm each took place at the right time and in the proper order. This issue of appropriate timing is key to growth and development in the body, as well. If every gene on a strand of DNA were triggered at the same instant, human development and growth would be impossibly chaotic. The activation of distinct genes at specific times during the period before birth and throughout life takes place in an elegant dance of impeccable timing.

For instance, between day eighteen and day thirty-four in the gestation of a human fetus, a host of special molecules, called transcription factors, bind to the DNA of the developing heart muscle and induce the genes to activate.[1] If the sequence of transcription factors is missing a component, or if any of these factors is dysfunctional, the chambers, valves, and associated vessels of the human heart will not develop properly, and the baby will be born with congenital heart problems.

Similar programs of gene activation are critical to the development of every part of the body, including the nervous system. As we grow toward adulthood, other transcription factors come into play, such as those that switch on particular genes in the male testes and female ovaries at the onset of puberty. Some of these molecules are also present in adults to help regulate normal activity in the brain.

]OOO[

LIKE OUR BIOLOGICAL LIVES, our personal lives unfold in a dynamic sequence of timed events. Signals and conditions change continually. Some changes we ignore; others demand our attention. Sometimes our dreams express themselves in discrete stages, with clear markers dividing the phases of life. Other times, we seem to transition smoothly and transparently from one stage of life to the next.

As a personal and professional coach, I often consult with people who need help with finding and activating dreams that will inspire the next phase of their life. Frequently, for instance, clients in midlife tell me that they have met the goals they set for themselves when they were young. Now, conditions have changed, and they feel at a loss. What's next, they wonder?

In our first coaching session, Victoria was clearly in distress. Ever since she could remember, she told me, she had wanted a large family. One of seven children, she loved the idea of raising her own. After two biological children, she and her husband, Mark, adopted three others. Now, her youngest daughter had gotten married and moved away from Portland. Though Victoria knew that holidays, visits, and the arrival of grandchildren would keep her close to her sons and daughters, her day-to-day life was no longer devoted to raising a family.

"I can't figure out what to do next," Victoria told me. "My mother filled her life with organizations and charity work, but that doesn't seem right for me. Mark's position at the bank provides us with a good income, so I don't need to work. I've joined a gym and signed up for yoga classes, but there are only so many hours you can spend on a treadmill. I feel I have lost my purpose for living. When I wake up in the morning, I feel so empty inside!"

"What did you study in college?" I asked.

"Philosophy," Victoria replied, adding with a rueful smile, "not exactly a practical choice." She went on to say that reading philosophy took her out of herself and gave her something "large and beautiful" to think about.

As our conversation continued, I explained to Victoria that we never have to find our way in life alone. Just as transcription factors help the DNA in cells to activate at the right time, there are always allies waiting to help us to find and activate our dreams. We mapped a strategy for Victoria to reach out to friends, including people she hadn't spoken to in many years. I suggested that she pay close attention to her conversations with them, as if each might contain coded clues pointing toward possible next steps.

During our next session several weeks later, Victoria told me about a talk she'd had with a friend who was a teacher. The talk had triggered a memory of how much Victoria had loved helping her children with their homework and volunteering at their various schools. I reminded Victoria that in searching for a dream, it is important to act on information as it is acquired. The end result of a series of steps may initially be hidden, but as we continue to take step after step, our goal starts to become clear.

After this session, Victoria decided to act on the information about herself that she had discovered during her conversation with her teacher friend. She made an appointment with the dean of the extension school at the local university. The dean was very approachable. He discussed the range of courses offered to adults and invited her to sit in on several classes.

Though the prospect of assignments, papers, and examinations seemed daunting after all these years, Victoria was able to overcome her fear by reminding herself that being in school had always made her happy. I love to read, she told herself, and class assignments will just point me toward many new and interesting books. After visiting several classes, she enrolled the next semester. Soon, she was deep into her studies, reading theories of history, analyzing the philosophies of various periods, and enjoying herself thoroughly.

After a year of refresher work, we began to formulate Victoria's next step. She set a goal to earn a master's degree at the university so that she would be qualified to teach the extension courses she had so much enjoyed. Her dream now is to help other adult students rediscover their love of learning and reconnect with academic disciplines they may have left behind years before. Energized by this dream, which she activated at just the right moment in her life, Victoria seems like a new person, filled with purpose and vitality.

FINDING AND ACTIVATING A DREAM

The questions in this exercise are drawn from coaching sessions such as those I engaged in with Victoria. They can assist you in discovering and activating a new dream.

Find a quiet place and take some time to reflect on the three questions below. They can help you discover whether, and to what extent, your life is still motivated by a dream. Stay with the feelings that emerge as you think about the questions. When your thoughts feel complete, record your answers in a journal.

[] What is the first thought you have when you awaken in the morning?

[] What is the first feeling you have when you awaken in the morning?

[] What is your expectation about how your day will go? About the relationships you will have today?

If your answers reveal that you have lost the sparkle and joy you had when you were motivated by a dream, it may be time to reach out for a new life goal. The first step in finding a new dream is to get back in touch with the feelings you had when you were unfolding a dream, such as those Victoria had when she was raising her family. Remember, as vividly as you can, a period in your life when your dreams were center stage and you were unfolding them in a fulfilling way. Now answer the three questions above as you might have answered them at that period in your life. Touch the feelings and allow them to emerge. Stay with your feelings until they are clear. When they feel complete, record what you have discovered.

With these feelings fresh in your mind and heart, answer the following questions to help you explore the seeds of a new dream:

[] What dreams for your life do you remember from your childhood? Which of these have you fulfilled? Which have been abandoned along the way?

[] What do you find easy to do and love doing?

[] Which of your talents, special interests, and hobbies make you especially happy?

[] What was your major in college? What about this subject particularly interested you? How have you put this discipline to use in your adult life?

[] What was your most enjoyable job? If you could design the ideal job for yourself, what would it be?

[] Look over your answers carefully. What similarities or patterns do you note?

Sometimes the process of finding a new dream requires a period of incubation. Ideas may emerge when you least expect them to. Before you go to sleep at night, or when talking to a friend or family member, be alert to clues that may spark a new idea. When an idea emerges, do not hesitate to explore it. Exploring an idea does not commit you to anything. Once you know the step you can take, do it! You may not yet understand where the next step is leading, but taking action, as Victoria did when she met with the dean of the extension program, may initiate a process leading to manifestation.

If, as happens for some people, you feel yourself unable or unwilling to take action, reflect on the answers to these additional questions:

[] Are you hesitating because you are afraid?

[] What fearful consequences do you imagine might follow if you take action?

[] Consider the possibility that underlying your fear is a limiting belief about yourself. What is the nature of that belief?

As you peel away the fear, allow yourself to feel the desire for change and the belief that change and growth are possible. Remind yourself that if you allow your fear to stop you, it will also stop you from enjoying many other aspects of your life. Don't run away from the fear or try to cover it up. Spend as much time as you need to articulate the limiting belief that is the source of your fear.

Now ask yourself:

[] What belief about yourself would support you in taking that next step?

]0000[

PLAY WITH TRANSFORMING your fear-based belief into a life-supporting belief. "I can't take a new direction at this stage in my life"—is an example of a fear-based belief. "I know more than I ever did; now I want to use what I know to develop a realistic plan to pursue my dream"—is an example of a life-supporting belief. Once you can articulate the life-supporting belief, take action. Action anchors the dream and begins the process of making your desire happen. One step leads to another. Soon, like Victoria, you may find your life expanding in new directions, infused with the joy of living a dream.

CHAPTER NOTES

1. Jonathan A. Epstein and Clayton A. Buck, "Transcriptional Regulation of Cardiac Development: Implications for Congenital Heart Disease and DiGeorge Syndrome," *Pediatric Research* 48, no. 6 (2000): 717–724.

CHAPTER 2

turn on and turn off

SOME PATTERNS ARE SO INTERWOVEN into the tapestry of life that evidence of them can be seen in every aspect of our physiology. Cells, organs, and systems may exhibit their own variations of a particular motif, but each expression is recognizable as an iteration of the same theme. Take, for example, the observation that everywhere in the body, times of intense activity alternate with times of quiet.

Taking Time to Turn Off

Everything in the body turns on and off periodically and repeatedly. This pattern is evident within individual cells, in interactions between cells, and in the cycles of organs and systems in the body.

Even neurons—the specialized cells of the nervous system that make it possible for us to breathe, walk, talk, think, see, taste, and feel—have off phases. Immediately after a neuron conducts an electrical impulse and releases chemicals from its ending, it turns off. During the off phase, called a refractory period, the neuron cannot be activated. It is unresponsive to any incoming stimulus, regardless of its strength.

If we were to observe a neuron during its refractory period, we might conclude that nothing is happening, because no electrical impulses are traveling across its membrane. Yet something very important is going on inside the neuron during its time off. In fact, the off phase is essential to the neuron's continuing ability to conduct impulses.

Impulses are transported across the membrane of an individual neuron by the movement of small charged particles, or ions. As sodium and potassium ions move into and out of the neuron, an electrical impulse travels across the membrane of the neuron and chemicals are released from the neuron's endings. These chemicals carry messages from one neuron to the next, along the chain of neurons that make up the nerve pathways of the body. After an impulse has been transmitted, the ions within each neuron must return to their starting points to be ready to conduct the next impulse. Only when the ions have had a chance to regroup, is the neuron ready to respond to the next stimulus.

]000[

LIKE NEURONS, WE HUMANS NEED REFRACTORY PERIODS in order to stay sensitive and responsive. Alternating active and quiet times is essential to living with exuberance. Ignoring the on-off rhythm that is part of life's natural template leads to exhaustion, loss of creativity, and troubled relationships.

Do you recognize yourself in the following scene?

Jane gets up before dawn, showers quickly, jumps on the subway, and walks briskly to her office at an investment firm. She's often there at seven, before anyone else is around. Immediately, she tackles her email and voicemail messages. Never present in the moment, she is always doing a second task while implementing the first. While she's answering her email, her mind is racing ahead to the voice messages, planning what she will say. Her attention is not focused on the email she is answering; she makes decisions quickly, keeping her responses short. She likes returning calls early in the morning because it takes less time to leave a voice mail message than to have an actual conversation. By eight, she's drafting a memo, sketching out the agenda for her morning staff meeting, and making notes in her planner. Multitasking is her middle name. In fact, she would consider herself inefficient if she weren't working on several things at once. Her appointments begin at nine and continue throughout the day. Generally, she eats lunch at her desk, allowing herself a restaurant meal only when there's a client to entertain. But long before the lunch hour rolls around, Jane is stressed and tired.

Sensitive? Responsive? Hardly. There's no time in Jane's busy schedule for her energies to regroup; no time off for renewal. Rushing from task to task without a break denies Jane the rest periods every living being needs to keep functioning at an optimal level. Like a neuron without a refractory period, Jane is likely to lose her ability to carry on her work.

What might Jane do? In the course of this book, I'll suggest many strategies for putting cellular wisdom to practical use. Here's a quick way to give the body/mind a refractory period in the midst of a busy day.

TAKING TIME OFF

Find a quiet place where you can be alone for ten to fifteen minutes— perhaps an empty conference room, a lounge, or your car if you're on the road. Decide to spend the next few minutes allowing your body to function naturally in a relaxed state. Articulating the intention focuses your energy.

Turn off the ringer of the phone and the lights. If music helps you relax, put on headphones and listen to a CD of pan flutes or other simple melodic sounds. You might also try a recording of ocean sounds or bird song. Take off your shoes and jacket; loosen your tie or belt. Sit on a comfortable chair, with your feet flat on the floor or, if you prefer, lie on the floor. Allow your eyes to close so that they rest as lightly as feathers.

Scan your body to locate the spots where you are experiencing the most tension. Does your head ache? Are your neck or shoulder muscles tight? Are your eyes tired? Does your lower back hurt? Send the thought to those areas that this is their refractory period, their time to relax and refresh.

Take three successive deep breaths, with your mouth closed. Take in each breath for three or four seconds, and then release it slowly for five or six seconds. During the out-breath, let go of the tension you have been experiencing in any part of your body. Utter a relaxing sound, such as "Aaaaahhhh" during the out-breath. Just allow the sound to escape from your body.

As you breathe in, muscles contract, your lungs inflate with air, and the hemoglobin of your blood is oxygenated so that your red blood cells can transport the oxygen through your capillaries to cells and tissues. Oxygen is fundamental to life—more basic than water or food. Oxygen fuels the cells so that they can transform nutrients into energy needed for the cells' vital activities.

As you breathe out, muscles relax, your lungs deflate, and the cells and tissues release waste products that they have metabolized. By resting and breathing deeply, you are helping your cells and tissues unclog and release these wastes.

After the first few deep breaths, allow your breathing to find its natural pace. Feel the air coming into your lungs. Imagine your cells and tissues being nourished. Feel the air going out of your lungs. Imagine your cells and tissues releasing anything they no longer need. The internal harmony, within and between cells and systems, will occur easily as you breathe. Allow this harmony of breath coming in and going out to lull you into quietness. Imagine that you are watching the waves coming into shore on the in-breath and flowing out to sea on the out-breath. Or imagine the life force of the universe entering your body on the in-breath, and your own life force expanding to the ends of the universe on the out-breath.

If a thought occurs to you, allow it to pass, without following it or giving it more attention. No struggle is necessary. Simply allow the thought to move out of your focus, like a cloud moving across a clear expanse of sky. Allow any images that appear to transform spontaneously. There is no need to control the images in any way. Just observe them changing.

After ten minutes or so, allow your body to begin to stir. Gently move your shoulders, stretch your muscles, and return to a normal waking state. Take a few minutes to wash your face or freshen your makeup and straighten your clothes. Treat yourself to an apple or a cup of tea. Make a date with your body to provide another refractory period at a later time. Then return, refreshed, to the tasks at hand.

Talking and Listening

Neurons communicate with each other primarily by chemical means. When a sending neuron conducts an impulse, it releases a chemical called a neurotransmitter. The receiving neuron has receptors that are perfectly matched to the neurotransmitter released by the sending neuron.

With each pulse of communication, a specific quantity of neurotransmitter is allowed to enter the space between neurons from small sacs where the chemical is stored. The release is no more and no less than is required to carry the message. The period between pulses of neurotransmitter release constitutes a "quiet" time and is critical to the entire process. Without intervals of quiet, information could not be coded properly for transmission from neuron to neuron.

Intervals of quiet are so essential that neurons have an additional mechanism that makes sure that the signaling between neurons stops periodically. Another group of chemicals called enzymes are present in the space between neurons. Enzymes break down neurotransmitters so that they can no longer activate receiving neurons. These enzymes destroy any leftover neurotransmitter that has not been taken up by a receiving neuron during transmission. Without these chemical bodyguards, receiving neurons would be on all the time. Soon they would lose their ability to take in new information.

]0000[

HUMAN COMMUNICATION ALSO HAS A NATURAL RHYTHM. Too often we forget that intervals of quiet are just as important to good communication as talking. Remember Jane, who rushes from task to task without ever taking a break? Because she talks to voice mail machines more often than to people, she seldom hears the unspoken messages in the pauses and cadences of conversation. Nonverbal cues, such as facial expressions, eye contact, and gestures, are also eliminated from her communication repertoire. During morning staff meeting, Jane speaks quickly, often interrupting or changing the subject. Is it any wonder that few people on Jane's staff really like her? With her communication button stuck in the talk position, Jane is considered to be a poor leader, in spite of all her hard work.

Jane's sister Martha is a family-practice physican in rural New Hampshire. Though her daily schedule is long and busy, her style of working and communicating is quite different from Jane's. Martha's day also begins at seven. But when she gets to the office, she does not immediately engage in outwardly directed activities. Instead, she spends an hour quietly reviewing her case notes on the patients she is scheduled to see that day.

Martha believes that interacting with her patients—listening to them talk about their health and their lives—is as important to her patients as her physical examinations and the results of any tests she may order. Her process of diagnosis often eludes definition. She honors her hunches and intuitions, and takes into consideration what a patient does not say, as well as what a patient does say.

One of Martha's patients, a fifty-two-year-old carpenter named Ralph, had recently been hospitalized for bronchial pneumonia. A year earlier, he had experienced a less serious bout of the same infection. After listening to his respiration and confirming that his lungs were clear, Martha invited Ralph into her office for a talk.

Martha and Ralph spent several minutes in a comfortable back and forth exchange about the weather, local politics, and the new home she was building in town. Martha asked Ralph how his wife and daughters were doing. In the long interval of silence that followed her question, she noted that Ralph looked at his hands rather than at her, that his breathing became shallow, and that he cleared his throat and coughed several times. Martha was careful not to interrupt the silence with another question or comment. She wanted to give Ralph the chance to receive and process his thoughts and then to respond.

In a hesitant voice, Ralph told her that he was very worried about his youngest daughter, Jolene, who had married a man Ralph did not trust. When Ralph had recently visited her, she looked strained and unhappy. He'd overheard Jolene telling her older sister that her husband often stayed out late and came home drunk.

"I know it's her life and her choice," Ralph said bitterly, "but she's still my little girl. Sometimes I get so angry I want to march right over there and give that husband of hers a piece of my mind. But I know better than to interfere, so I just keep my mouth shut."

As Ralph was speaking, Martha was listening carefully, both to what he said and to what he did not say. She surmised that Ralph's wife and two daughters often excluded Ralph from discussions about their problems for fear of arousing his anger. This information gave Martha a clue to his pattern of illness. She knew from experience that unexpressed grief and anger can contribute to asthma or other lung problems. She took out her prescription pad and wrote on it: Tuesdays. 8 P.M. Sixth and Maple.

"What's this?" Ralph asked.

"It's when the men's group meets at the church uptown," Martha replied. "I hear good things about the meetings from many of my patients. I think your health will improve if you have a group of fellows you can talk to and listen to regularly."

"Well, that's the strangest prescription I ever heard," Ralph said as he shook Martha's hand. "But you've been real kind to listen to me blather on about my troubles, and I do feel better. Maybe I will check out that group."

Good listening helps Martha be a better doctor. Communication does not consist of simply transmitting information. Intervals of silence in any conversation allow both sender and receiver to process the information that has been transmitted and to regroup for the next exchange.

Balancing Acts

It makes sense that neurons connected to each other through circuits pulse together. Some circuits pulse intrinsically, with no input from other circuits. If these circuits are placed in a culture dish, they will continue to pulse as a unit.[1] Other circuits coordinate their actions with another division of the nervous system to achieve synchrony.

Let's look at one such balancing act within the body's autonomic nervous system.

The job of the autonomic nervous system is to control the motor functions of the heart and blood vessels, lungs, intestines, glands, and other internal organs. The autonomic nervous system has two components, the sympathetic nervous system (SNS) and the parasympathetic nervous system (PNS).

The SNS mobilizes energy. You might compare it to the gas pedal of a car; it makes the engine go faster. The SNS prepares the body for fight-or-flight by mobilizing the body's energy reserves to increase the rate and force of heart muscle contractions; raise the blood pressure; and increase the supply of blood to the muscles of the arms, legs, lungs, heart, and brain.

By contrast, the PNS is like a car's brake pedal. It minimizes activity, slowing everything down. PNS decreases blood flow to the heart and lungs, which slows the heart rate and respiration to rest these vital organs. It also restores the body's energy reserves by maximizing the absorption of nutrients from the intestines.

These two components of the autonomic nervous system operate in sync to maintain balance in the organs of the body. Normally, each period of intense activity generated by the SNS is followed by a period of restoration of body resources mediated by the PNS.

The SNS is called into action whenever we interpret a situation as threatening. For instance, if you expect that the person you are telephoning may deliver bad news, your palms may sweat and your heart may race as soon as you start dialing the number. You may experience the same symptoms in a doctor's office if you are worried about the exam. Even walking toward your boss's office, if you expect she might give you painful feedback, can cause a physiological response.

But any machine left in the on mode for long periods of time eventually wears out or breaks down. The human body is no exception. When the flight-or-flight mechanism of the SNS is stuck in the on position, your body's energy reserves get depleted, and you're on the fast track to stress-related dysfunction, such as heart disease, migraines, hypertension, and depression.

To understand why this is so, we need to understand something about how the human nervous system functions. The nervous system is actually an elegantly complex, interconnected network of nerve cells, or neurons and neural pathways. Only a small fraction of the body's nerve cells—about 10 percent—sense and bring in information from the outside world and control body movements; the other 90 percent communicate along pathways within the body, to internal organs and organ systems. The job of this neuronal network is to make sense of what we see, hear, taste, touch, and feel, and to make choices about what we do, and when and how we do it.

The SNS and PNS are integrated within this complex network. When we experience a dangerous or stressful situation, the SNS communicates with target organs, such as the heart, by secreting a neurotransmitter. The PNS secretes a different neurotransmitter. These two neurotransmitters have reciprocal effects: the SNS gives the heart a "go" signal, telling it to beat faster and pump more blood; the PNS sends the opposite message. Internal sensors embedded in blood vessels, close to the heart, send information about how hard the heart is pumping to the region of the brain called the hypothalamus. The hypothalamus then sends signals through the PNS or SNS systems to adjust heart function accordingly.

But in addition to stimulating the heart directly, the SNS also controls the adrenal gland, located just on top of the kidneys. When secreted by the adrenal gland, a neurotransmitter, identical to that secreted by the SNS, is carried in the bloodstream throughout the body. In that way, the adrenals cooperate with the SNS by speeding up the heart and inducing the flight-or-fight response. When the secretion of the neurotransmitters slows, the heart slows down.

The story is more complex yet. When we face stress or experience stress, the hypothalamus and the pituitary gland also send signals to the adrenal gland to secrete a steroid hormone called cortisol. One function of cortisol is making glucose available to the brain to help us cope with stress. Without cortisol, we would die at the first sign of stress. But too much cortisol, such as results from prolonged stress, causes high blood pressure and several other diseases.

Notice that the same structure within the brain, the hypothalamus, mediates both the functioning of the autonomic nervous system (PNS, SNS, and the neurotransmitter secretion from the adrenal gland) and the functioning of the endocrine system (cortisol secretion from the adrenal gland). Moreover, as part of the great neural network, the hypothalamus exchanges information with the part of the brain that processes emotion, the limbic system.

Given the functional linkage between these three systems—autonomic, endocrine, and limbic—it is not surprising that emotions alter the functioning of organs throughout our bodies. Nor should it surprise us that we can balance the activity of the SNS and the PNS by relaxing, meditating, listening to music, or engaging in other activities that improve our emotional well-being.

How many times in your day does your body experience a flight-or-fight response? Does an imbalance between the SNS and the PNS threaten your physical health and emotional well-being? Here are two quick exercises to help you assess your SNS–PNS balance.

A BIGGER PIECE OF PIE

Draw two circles, side-by-side, on a sheet of paper. Divide the circle on the left into sections, as you would slice a pie, to indicate things that are important to you. The size of the slice equals the degree of importance. Divide the circle on the right into slices that reflect how you spend your time. Again, make the size of the slice approximate the amount of time spent on each activity. Compare the circles, and see what you notice.

[] How you spend most of your time?

[] What are you neglecting that is important to you?

[] How much time do you spend on physical exercise?

[] How much time do you spend alone, reading, thinking, or just relaxing?

FIGHT OR FLIGHT INVENTORY

Think through your answers to the following questions:

[] What is your typical response to waiting in line at the bank, the supermarket, the gas station, or in traffic?

[] How do you typically react to a canceled flight or a long delay at the airport? How do you typically react when a coworker, friend, or family member fails to do something you were counting on?

[] How do you react when the telephone interrupts something you are doing?

[] How much alcohol do you drink each day?

[] How often do you speak to or visit your friends?

[] When was the last time you wrote a personal letter?

[] How often have you been angry this week?

[] How many hours of sleep do you get on a typical night?

[] When was the last time you had a check-up?

[] When was the last time you had your teeth cleaned?

[] Do you have any recurring health problems?

[] How many times have other obligations caused you to miss family activities?

[] When was the last time you told someone you loved them?

[] How often do you speak or write about what bothers you?

Review your responses to these exercises. You may also wish to discuss your responses with a friend or family member to get an outside perspective. Is your life in balance? What might you do to correct any problems you note?

]000[

CELLULAR WISDOM TELLS US that we can help to re-balance our lives by finding ways to express emotions, promote feelings of love and warmth, and avoid negative thinking. Each of these re-balancers has been shown to reduce variability in the heart rate. Talking or writing about your feelings, listening to music, and engaging in regular physical exercise are three easy ways to boost PNS response.

Also, explore your temporal environment for innovative ways to use interrupters or other signals, such as the telephone. When you hear the phone ring, rather than rushing to answer it, deliberately slow your pace. Move toward the phone without hurry and take a deep breath before picking up the receiver. Take the few seconds you've gained to center yourself, prepare to interact with the caller, and reaffirm your intention to speak authentically and clearly.

Hidden periods of time, such as the time you spend waiting in lines, can also be openings to a larger perspective. Rather than allowing yourself to get frustrated, welcome such times as opportunities, and explore ways to use them creatively. Use the time to invent a scheme or make a plan to do something just for fun. Use the time to dream, to connect, to center, or to create.

Here's a technique that works for me. Whenever I come across a quotation that inspires me, I write it on a small slip of paper and put it into my jacket pocket. When I find myself in one of these hidden periods of time, like waiting in line at the bank, I pull out a slip of paper and allow myself to expand into a larger vision by reading the quotation.

Reciprocity, like the balanced functioning of the SNS and PNS, can also be achieved by working with a partner who balances your individual style. Often, the most productive teams are made up of individuals with different creative styles or skills. Mary Lee, for example, is an organized, linear thinker. Detail-minded, she talks fast, solves problems quickly and efficiently, and enjoys tracking of each phase of a project as it unfolds step-by-step. Mark is her partner in a catering business. He is a creative visionary who would rather look at the big picture. Details clutter his thinking, but he's great at imagining how new projects will turn out.

Usually Mark makes the presentations to potential clients. He is an engaging speaker and paints a vibrant verbal picture of the glorious wedding or corporate party his company can arrange. When the prospective client begins to ask questions about details, Mary Lee takes over and outlines the timeline and planning steps necessary to bring Mark's creative vision to completion. During the rest of the presentation, Mary Lee and Mark defer to each other's particular expertise and style to best answer the prospective client's questions. One might say that their synchrony resembles a pulsating SNS–PNS rhythm.

Cycles

The monthly cycles of the reproductive system are controlled by an intricate pattern of signals between three organs—the brain, the pituitary gland, and the two ovaries—and three whole body systems—the central nervous system, the endocrine or hormonal system, and the reproductive system. The precise timing and coordination between these parts of the body demonstrate the critical role cycles play in cellular wisdom.

Reproduction depends on the ovary providing an ovum, or egg, that can be fertilized. A complex sequence of signals between the brain, the pituitary gland, and the ovary ensures that an egg is expelled from the ovary at the appropriate time. The cycle begins when neurons in the brain secrete a small molecule, a hormone called luteinizing hormone releasing hormone (LHRH). This hormone is released in a pattern of pulses. The pituitary gland receives these pulses from the hypothalamus. It then secretes its hormone, luteinizing hormone (LH) in a pattern of pulses. These pulses of LH signal the ovary to release an egg.

The complete cycle of signals between the hypothalamus, the pituitary, and the ovary varies in length from species to species. Rats, for instance, have a very short cycle, only four days. Hibernating bats ovulate only once a year. Humans, typically, have a monthly cycle, with the pattern of signals taking place every twenty-eight days.

Scientists researching infertility in women have discovered that intervals of quiet, when no signals are being sent, are as important to normal reproduction as the sending of this complex series of pulse signals. Doctors treating women who were unable to ovulate tried stimulating the ovary to release an egg by giving the women a continual supply of LHRH. They found, however, that continual stimulation of the pituitary by LHRH actually causes the gland to stop responding. Eventually, the pituitary loses sensitivity and becomes incapable of responding at all. They discovered that the effective way to stimulate ovulation was to give women the hormone periodically, in a manner that mimics normally occurring pulses.

The final signals in the reproductive cycle come from the ovary itself. Cells remaining after ovulation secrete the hormone progesterone, which signals the hypothalamus in the brain that ovulation has taken place. Neurons in the hypothalamus do not secrete LHRH as long as progesterone is present. Over the next fourteen days or so, the level of progesterone decreases, and the ovary begins to develop another egg follicle for the next ovulation cycle. The newly developing follicle secretes the hormone estrogen, which signals the neurons in the hypothalamus to begin to synthesize and secrete LHRH. Then the cycle begins again. Without times of quiet, the principle players in the beautiful symphony of new life would not be able to hear its intricate melody.

]00O[

JUST AS THE CYCLE OF REPRODUCTION must proceed through a harmonious sequence of stimuli and responses, so, too, our creative life unfolds with alternating periods of insight, or breakthrough, and seeming quiescence. At times when it seems that not much is going on, much may be happening beneath the surface to prepare the ground for the next burst of creative activity.

I think, in this regard, of my friend Elsie, a vibrant, energetic woman in her early forties. After spending ten years at home raising her three children, Elsie returned to the job she had held before her kids were born, as a legal secretary. Though she easily brushed up on her office skills, and she found the personnel changes at the law firm easy to get used to, Elsie began to feel frustrated after several months on the job. The work, which had seemed so absorbing ten years before, now seemed routine. Generating one contract or legal document after another now struck her as mechanical and boring. The job had not changed in ten years, but Elsie had.

One night, to cheer her up, Elsie's friend Susan invited her to a writing group that met in the home of a mutual friend. It soon became clear that the women in the group were not professional writers. One was a real estate broker who wrote poetry. Another was a programmer at a telecommunications firm, who was writing a television screenplay. A pharmacist was working on a novel. Each week, the women met to read each other's work in progress and to offer suggestions and support.

Though Elsie just listened that first night, a process began in her. She had always loved stories. Her favorite time with her kids was bedtime. Rather than reading to them, she would tell them stories she made up—an interconnected weave of fantastic adventures set in an imaginary world. Elsie had never considered writing down her stories. Making them up was just something she did for fun.

It now seemed to Elsie that she had been waiting quietly for a signal, something to tell her what to do next. She joined the writing group and began to write her stories, hesitantly at first, but then with increasing confidence. She found that her routine work at the law firm gave her plenty of time to think through the plot of the story she was writing, or imagine the details of a character or scene.

The group provided Elsie with a new audience for her stories. The other women loved the tales of adventure she spun and offered much support and many useful suggestions. One night, the editor of a monthly children's literary magazine was attending the group's meeting. After hearing the latest installment of Elsie's unfolding adventure, she asked whether Elsie would consider submitting her stories for serialization in the journal. Elsie agreed, with pleasure.

The cycle of Elsie's life had come full circle. Her quiet years at home had, in fact, been a time of creative gestation. Even the boring routine at the law firm played a role in preparing Elsie for the next stage of her creative life. Just as a field needs occasional fallow periods in order to be productive, our lives need similar periods of routine and introspection to allow us to hear the signals that point toward the next stage of our creative unfolding.

Body Rhythms

Many biological rhythms are circadian; that is, they occur every twenty-four hours. Body temperature, blood pressure, blood sugar, hemoglobin, and hormone levels rise and fall in a regular day-and-night rhythm. We've all experienced how important these rhythms are to good functioning. When travel upsets our normal rhythm, we call it jet lag and blame it for our sleep disturbances and diminished ability to think and respond clearly.

Other waxing and waning rhythms occur monthly, seasonally, and annually. As we have seen, the monthly cycle of human reproduction is an obvious example of this pattern, as is the seasonal hibernation of bears and other mammals. You might even say that, viewed as a whole, our lives are marked by a sequence of timed events of even longer duration, which we call infancy, childhood, adolescence, young adulthood, middle age, progressing into our elder years. The periods of these rhythms vary, yet the pattern is a basic, repeating, dynamic theme of all life. Rhythms of different frequencies may come in and out of phase with each other, yet there is an overall, resonating harmony that embraces all variations.

As we have seen, too many of us ignore the periods of quiet or rest that are a necessary component of each of these rhythms. We pattern our lives with extended periods of activity, allowing little or no respite for weeks, months, or years. Cellular wisdom teaches us that active and passive stages must alternate for systems at all levels to function well. The balance point may differ according to our personalities and styles, yet the basic pattern cannot be ignored without risking harm to ourselves or diminishing our effectiveness. Our creativity, the quality of our relationships, and the maturation of many processes is enhanced by such frequent periods of quiet. We simply think more clearly in the free space of an uncluttered mind.

Many of us have experienced, for instance, that important insights about a project we've been working on feverishly all day come when we take a break and go for a walk, or when we wake up refreshed from a good night's sleep. Pauses induce equilibrium and a feeling of being in touch with ourselves. They give us an opportunity to let go and experience a sense of release.

Letting go means putting away our worries and the sense of responsibility that so often weighs us down. Worry makes creative thinking more difficult. It blinds us to multiple perspectives and alternatives. Letting go, by contrast, allows us to relax into the natural rhythm of what is trying to unfold. It opens the mind and gives us access to knowledge we didn't know we had. Letting go takes only a moment, yet it restores our balance, creativity, exuberance, and passion for life.

As we relax the grip of control and allow the waxing and waning of influences to stream to us, it may often seem as if the universe conspires with our essential selves to push us along toward the next step or stage in life. I believe that by clarifying our desires, focusing our energy on taking the next steps and anticipating the outcome that we envision, we can create situations that are favorable to these influences. Engaging in these activities signal that our energy is in tune with the energies of the universe, and invites a response.

When we are in rhythm, synchronicities abound. A chain of events leads to a new career as a writer, as happened to my friend Elsie. We hear an interview on the radio or pick up a book that points us toward the answer to a long-held question. A chance occurrence catalyzes a breakthrough that allows us to understand a problem we've been working on in an entirely new way. These harmonic convergences between ourselves and the larger universe occur only if we create the time and space for the natural waxing and waning of life energy to follow its own pattern. Our job is to remove the constraints of stress, over-activity, and worry so that the energetic patterns can move through us without blockage or disturbance. When we trust the process and allow ourselves to relax into our natural rhythm, life unfolds, effortlessly.

Why Learn the Hard Way?

I learned the hard way that activity and rest must be balanced. Perhaps, like me, you've allowed your life to be overwhelmed by continuous activity. When you are unable to let go and relax into periods of stillness, your body will often intervene to remind you to do so. The pace of my last years in academia took their toll. A serious illness forced me to stop work for several months to give my body/mind a chance to catch up.

Why wait for illness to disrupt your life and force you to take the time outs that cellular wisdom teaches us are essential? Why risk losing your passion for your job, your relationship, and your interests by doing too much? Why wait until emotional and physical exhaustion force you to turn off or pull the plug?

When illness forced me to rest, I had time to evaluate how I had been living. I realized that I had ignored things that were really important in order to meet what I thought were my duties and obligations. I had neglected my family, my health, my spiritual life. In essence, I had abandoned myself. My life was out of control, with no balance between action and quiescence.

I have spent the past several years learning again how to take time to turn off, to balance talking with listening, and times of high activity with times for the replenishment of resources. I have worked hard to allow my cycles to unfold and to get back in touch with my natural body rhythms. In short, I have learned how to live.

After my sabbatical and leave of absence, I restructured my life to alternate periods of activity with quiet times for intensive reflection. I generally spend the first three hours of the morning working quietly by myself—reading, writing, and reflecting without interruption. I schedule appointments with my coaching clients into blocks, with time off between the blocks to make notes and regroup. I try to block out one day each week for writing, scheduling no appointments or other outside obligations for those days.

I also allow my body to find its natural daily rhythm. Some mornings I awake very early with new ideas about something that I have been writing or a way to help a particular client. I always honor those periods of particular clarity, allowing the ideas to flow through me and for connections to emerge without blocking or censoring what emerges. When I am not traveling to a workshop or conference, I allow time to be in the natural surrounding of our Colorado home, realizing that a change of scenery is important to continuing a creative process. Days at home in the country always include a lot of reading, writing, and being with our two cats.

Take a moment to take stock of your own daily, weekly, seasonal, or annual patterns. Ask yourself what you can do to create the time and space for quiet periods each day, each week, and each year. Realize that times of quiet are wells of nourishment that can bring insight and harmony into your life, and make a commitment to explore ways to bring these periods into your life. Be creative, insightful, and imaginative in considering how you can do this. Doing so can rekindle your passion for life.

Honoring the on-off dynamics of cellular wisdom will allow you to mobilize your life energies to create, relate, and live exuberantly. Only then can you resonate with the vibrant pulsing of life.

CHAPTER NOTES

1. Robert Y. Moore MD, "Circadian Rhythms: Basic Neurobiology and Clinical Applications," *Annual Review of Medicine* 48, no. 1 (1997): 253–266.

CHAPTER 3

shape life moment-by-moment

THOUGH WE MAY BE AWARE of exterior changes in our bodies—a few extra pounds or gray hairs, better flexibility after six months of yoga classes—we are seldom aware of much more significant changes that are going on inside our organs and body systems. Though women may be quite aware of the physical changes that accompany their monthly reproductive cycle, or those changes brought about by pregnancy and childbirth, the organs and body systems of both men and women are changing all the time in response to internal and external triggers. This capacity of the body to change dynamically is known as plasticity.

Neuroscientists like myself are most excited by recent evidence that the structure and physiology of the adult brain is continually changing. Plasticity in the brain includes physical changes in the dendrites through which neurons receive information, and in the axon terminals through which they transmit information to other neurons. Most dramatic, however, is the finding that new neurons are continually being born in the adult brain.[1] The physical changes in the brain that we explore in this chapter have exciting implications for learning and growth throughout life.

Reproductive Signals and Responses

The monthly changes in the female reproductive system are the most familiar example of plasticity in the body, so let's start by taking a closer look at what happens to trigger the changes women experience. We've talked previously about the fact that in addition to neurotransmitters that carry messages between cells, neurons also secrete hormones. In the introduction, I described my laboratory research involving neurons in the brains of female rats that secrete a hormone (LHRH) that controls the pituitary gland. In response to that hormone secretion, cells in the pituitary gland secrete another hormone (LH) into the bloodstream. When LH reaches the ovary, it helps bring about ovulation, the release of the ovum or egg that begins the fertile phase of the female reproductive cycle.

Timing is everything. Each step in the complex interplay of hormonal signals and responses is critical to reproductive success in all mammals, including humans. The initial "go" signal from the hypothalamus must be sent at the right moment, exactly when the follicle in the ovary is mature and the egg is ready for release. If the egg meets the sperm and is fertilized, it implants into the wall of the uterus and begins developing into a fetus. If it does not, the lining of the uterus, which had thickened in preparation for implantation, breaks down, and menstruation occurs. This interplay of signals and responses occurs month after month throughout a woman's reproductive life, which can last up to forty years.

]0000[

THE COMPLEX SERIES OF CYCLICAL CHANGES can teach us much about bringing about significant change in our lives. For instance, how many times have you encountered a barrier and been stopped cold? Have you tried to give up smoking or change another habit, only to fail? The message from the body is that change is possible. Life is shaped moment-by-moment. If the timing isn't optimal, you can wait for the right set of circumstances to emerge. You can choose to move to new internal environment, taking advantage of life's dynamic character. You can ask, "What is the trigger that's needed to bring about the transformation, to break through the barrier?"

An example from my own experience illustrates this process. When I arrived at Tufts University as an assistant professor, I had a very small office, about ten feet by ten feet, with no windows. I was smoking at the time, and I smoked more when I was under stress. When I was writing a grant application, I usually smoked about a pack of cigarettes a day.

One morning, after a twenty-hour day spent working on a grant, I opened my office door and was almost bowled over by the smell of stale cigarettes. My stomach churned and my heart was pounding. I had quit smoking many times, only to begin again a month or two later. But standing there at the door of my office, I promised myself that when the grant application was mailed, I would not smoke again. That was November 1981. I mailed the grant, and I have not smoked since!

What was different this time? Why was I able to break through? For one thing, the smell of stale cigarettes in my office reminded me powerfully of what I already knew but was choosing to ignore: I, too, smelled like cigarettes. My clothes, my hair, my home, my breath—they stank like my stale office! My husband often pointed this out. He had even threatened to stop kissing me if I continued to smoke. My gut reaction to the smell of my office was a trigger, a signal sent from deep within me that the time had come for me to quit. Under the strong influence of this signal, I felt a sense of urgency to make a change. The timing was right, and I was primed to move to a new environment.

From this experience and my understanding of body processes, such as the sequence of hormonal signals and responses I had been studying in laboratory, I learned that functional change follows a pattern of five steps:

1. Identify a change you wish to make.

2. Allow yourself to experience internal triggers that motivate change.

3. Make sure the timing is right.

4. Identify the barriers that need to be broken through.

5. Choose a signal to start the process.

My internal triggers for stopping smoking were my revulsion at the smell of stale cigarettes in my office and on my clothes, and my unhappiness over my husband's negative response to my smoking. Moreover, the timing was right. I was really ready to quit. The barriers that needed to be broken through were my long-standing, habitual patterns, including the pattern of quitting many times and starting up again. The signal I chose to set the process in motion was mailing in the grant application I had written.

When I left academia, I gave up drinking coffee. I loved espresso and cappuccino, and I had a case of special coffee, Cafe du Monde, shipped to me from New Orleans every month. Now, given the functional change in my life, I was determined to know each day that I was living in a different style, without the pressure that comes from academic life. As a symbol of this change, I chose to eliminate coffee. I have not had a cup of coffee since September 1997. Again, the time was right to make this change. In breaking away from my habit of drinking coffee, I was consciously creating a new environment that would nourish me intellectually, emotionally, and spiritually.

Ask yourself: What changes do I want to make in my life? What new environment do I want to create?

Catching a glimpse of how life could be, can help motivate you to make a functional change. In my own case, I enjoy living in a smoke-free home. No longer do I find cigarette burns in the folds of my clothes. I awaken from sleep without the foul taste of lingering cigarettes. My teeth are white instead of yellow. My fingers are free of cigarette stains. Our car is free of cigarette butts and cigarette burns. The quality of my life and my health is improved. Having quit drinking coffee, as well, I feel more relaxed, less on edge, generally healthier.

Ask yourself, What is the trigger for me to change?

For some people, especially those struggling with addictions, the trigger is a desperate desire to reclaim their lives. Sometimes, you need only to admit an unwillingness to continue in a unhappy state, and to say to yourself, I will not live like this any longer. When my coaching clients want to make a change and yet fail to do so, I ask them, "Haven't you suffered enough?" When you reach a point where you can no longer tolerate the consequences of your behavior, you create a new internal state and trigger a set of events not unlike the complex, interrelated signals and responses that make reproduction possible.

Is there a right time to make these decisions? If you want to make a change and cannot do so now, can you commit to doing so at a time that is right? Though I made the decision to stop smoking spontaneously, I did not quit immediately. My sense of timing indicated that my level of stress would be much lower, and that I would have a better chance of success after I mailed the grant. Finishing that project was the signal I chose to set my change in motion. Similarly, I stopped drinking coffee when I left the academic world. Because I wanted to sustain a slower, more mindful pace in my new life, I chose to give up coffee on the first day of my new life after leaving Tufts.

Have you reached the point of choice? If not, what are the barriers to your making these choices? How can you break through those barriers?

Thinking back, I realize that my previous attempts to quit smoking had been half-hearted. Though I knew I *should* stop, part of me rationalized the habit as a way to control my weight or burn off nervous energy. Long-standing habits are hard to break. At some level of our being, the habit comforts us and fulfills a need; otherwise, we wouldn't be doing it. A first step to breaking through is to become more mindful about the behavior you wish to change. Rather than acting automatically, make each instance of engaging in the habit a conscious choice. Be on the lookout, all the while, for signals that you are ready to make a change.

Breaking Through to Synchrony

The body changes we've been looking at so far in this chapter are triggered by the release of hormones and other internal signals. But the body can also change in response to external triggers. For example, when a baby starts to suckle a mother's breast, a complex series of internal changes in the mother's body are set in motion.

As the baby sucks the nipple, nerve endings in the mother's breast are stimulated. The signal is transmitted, via nerve tracts, up the spinal cord to neurons in the hypothalamus region of the mother's brain. The message is received by a specific group of neuroendocrine cells. These neurons produce a specific hormone that is responsible for inducing milk to let down from glands in the mother's breasts. As the baby suckles, milk is ejected and flows to the nipple.

Though this process may seem straightforward, here, too, a barrier must be broken through. Between feedings, when the baby is not nursing, small projections of cells that are not neurons separate the bodies of the hormone-producing neurons from each other.[2] When the signals from the nerve endings in the breast reach the neurons in the hypothalamus, these projections are retracted so that neurons producing the hormone can contact each other with no separation. This contact allows the neurons to synchronize their action. The neurons fire in synchrony and release a large amount of hormone, producing the flow of milk from the mother's breast in response to the baby's suckling.

Sending Signals to Others

Signals from outside can affect us psychologically and emotionally, as well as physically. Depending on how we respond to them, such signals can either set up barriers that block our access to success and happiness or help us break through the blockages that keep us from achieving our goals.

Brianna, for instance, always had trouble in school. The dyslexia that hindered her ability to read and do math had never been properly diagnosed or treated. As a result, Brianna tended to keep quiet, as other students often laughed when she reversed letters when she read or mixed up the numbers in math problems. She never volunteered an answer to a teacher's question, for fear that she would embarrass herself. As her high school graduation approached, she was inwardly elated. No more school, she told herself. No more embarrassment.

However, there was one kind of lesson Brianna loved: the dance class offered as an after-school activity. When Brianna was dancing, she entered another world, a world where she belonged. It was as if there were no barrier between herself and the music, and her body flowed in synchrony with the other dancers.

Brianna could not convey to her parents how she felt when she danced. Though her dance instructor urged Brianna to continue to study dance, her parents were not encouraging. They wanted Brianna to settle down and find a steady job, perhaps at the clothes factory in town. Luckily, the "stop" signals being sent by her parents were countered by "go" signals from her dance teacher. The teacher helped Brianna find a job in a dance studio in the city. In addition to the small salary she received for cleaning and answering the phone, Brianna was allowed to attend one dance class each day.

In this new environment, far from the negative signals she had been receiving from her parents and schoolmates, Brianna began to blossom. She found herself more and more at ease with herself and more and more comfortable with others. She began to feel at home, wherever she was. Younger dancers at the school came to Brianna for help, and she soon began assisting in teaching beginners. Whenever she worked with a young dancer who seemed painfully shy and silent in conversation, but who lit up when the music started, Brianna thought of her own experience and made a special effort to send the young dancer strong signals of approval and support.

Like the neurons in the hypothalamus, we are often functionally separated from other people who could act along with us in synchronous ways and, in so doing, increase our ability to bring about change. Peer groups that work together to support each other in making functional changes are a much more effective way to send and receive helpful signals than trying to do it alone. For example, members of various change communities, including 12-step groups, personal growth workshops, Internet communities, church and meditation groups, and artists' and writers' circles, often come together to fire in synchrony so as to bring about powerful and lasting changes in their lives.

Note that sending a signal by acting in a synchronous way is quite different from urging someone else to make a change. Few people successfully stop drinking alcohol or permanently lose weight simply because someone else urged them to do it.

Think about changes you wish to make in your life, and ask yourself the following questions:

- Are there others who would like to make similar changes to the one you wish to make?

- How can you come into closer contact with these people so as to send each other helpful and supportive signals for change?

After leaving academia, I felt a strong need for such a support community. I began attending a series of human potentials workshops led by a teacher who catalyzed personal growth. The synergy of this group of people who were transitioning to new lives, opened portals of my own way of thinking and being, and formed the foundation for the explorations that I have engaged in since that time, including writing this book.

New Pathways for Learning

The stimulation we receive from attending a personal growth workshop, as well as any other kind of novel and complex experience, causes physical changes within the brain—further evidence of the brain's plasticity. In one part of the brain associated with learning, the hippocampus, the number of inputs to a neuron increases in relation to the degree of external stimulation we experience. Reading a complex novel, planning a vacation, taking a class in Indian cooking, or trading options—anything that brings something new and multifaceted into our lives—changes the neurons of the hippocampus so that they can receive more information and form new, more complex pathways to other parts of the brain.

Evidence of the importance of the hippocampus in memory and learning comes from studies of animals and people following traumatic injuries or surgery. A famous case reported in the 1950s describes a patient identified by the initials H. M.[3] This patient underwent surgery to remove portions of the temporal lobe on both sides of the brain in an attempt to alleviate his severe epilepsy, which originates in a dysfunctional hippocampus. Because the surgery destroyed the hippocampus, H. M. lost the ability to learn and remember new events. He would forget that he had recently eaten and attempt to eat again. He could not learn the names of people he met, or remember who they were.

Today, after almost fifty years of intensive research and the emergence of new technologies, such as magnetic resonance imaging (MRIs), we still do not know exactly how learning and memory occur.[4] We have established, however, that learning and memory require chemical changes in the hippocampus and the creation of new pathways between the hippocampus and many other of the brain's more than ten billion neurons.

The receiving components of neurons are primarily their dendrites. Dendrites resemble the branching limbs of a large tree. Along the branches of the dendrites are short protrusions called spines, which are positioned like the thorns on the stem of a rose. Unlike rose thorns, however, the spines are rounded rather than pointy, shaped somewhat like mushroom caps. Most of the connections that bring input from other neurons to the neurons of the hippocampus make contact on these spines, which increase in number and change their shape rapidly when we are engaged in complex, stimulating activity.[5]

Present in the now, neurons are continually aware of our changing circumstances. Moment-by-moment, they are alert and ready to respond to increases in stimulation by creating new circuits and transferring new learning to other parts of the brain.

Pathways in the Emotional Brain

In addition to its function in learning, the hippocampus is a component of the emotional brain. Circuits routed through the hippocampus allow emotions to build, accelerate, and amplify. Often these circuits fire in response to a memory.

Do you laugh when you see your old high school yearbook photo? The hairstyle and clothes seem so strange, you can't imagine wearing them. Yet there's your picture, proof that you did. Other memories are not so funny. Perhaps there is someone who has hurt you or wronged you, someone you have never been able to forgive. When you recall that person or an incident that caused you to suffer, do you feel anger, resentment, emotional pain?

Early in evolution, single-cell organisms could sense elements in the environment and respond to them, though every stimulus and response was processed within the single cell. As more complex organisms evolved, one group of specialized cells took over the sensory function of processing incoming information, and another group took over the motor function of responding to this input.

The greatest evolutionary strides were made as the human nervous system developed. While 20 percent of the neurons in our bodies are still involved in receiving sensory stimulation and responding to it with movement, more than 80 percent of the nervous system forms a great interconnecting neural web. This complex weave of pathways and connections allows us the greatest flexibility in perceiving and combining various kinds of stimulation. More pathways mean more alternatives and a greater ability to choose a response from a vast repertoire of behaviors.

Current thinking holds that when neurons are firing in one specific pathway, they cannot be induced to become part of another pathway. The implications of this neurological supposition are profound: If we choose to engage in remembering old events or rehearsing the circumstances of old pain, the neurons of the hippocampus fire continually to bring those memories and the associated emotions to awareness. Thus, they are not available to respond to stimulation and develop pathways for new learning.

]OOO[

WE EACH CAN THINK OF OCCASIONS or sets of circumstances in our lives in which we have felt caught in a dynamic of emotions that seem as if they will never change. I grew up in a first-generation Italian family on my father's side. Every Sunday my parents and I were expected to be present for a family dinner. Food was plentiful, as was conversation between the seven brothers and five sisters on this side of the family, their spouses and children, and my grandfather. My grandmother was always in the kitchen.

Decisions about business were a prominent topic of conversation between the brothers and the brothers-in-law. Not so openly discussed were the many family interpersonal intrigues. I soon learned who was speaking and not speaking to whom, and whose name was not to be spoken aloud in the house. A daughter could be disowned if she married a man of whom her father did not approve. The name of this daughter could never be spoken in the house, and she would be prevented from seeing her mother. I heard mothers pleading to see their daughters, and knew of daughters who were so angry they vowed never again to set foot in the family home. As I grew older, I came to understand how much pain these long-held resentments inflicted on everyone involved.

Keeping such destructive patterns active by continually renewing past pains uses neurological energy and resources. When part of our neural net is diverted to such unproductive activity, we are thwarting the enormous potential of our cells to learn and so shape our lives toward new growth. Taking a lesson from the plasticity that characterizes our physical bodies, we can choose to respond to old pain in new, more creative ways. When we do so, we allow our brains to forge new psychological and emotional pathways. Take the risk, forgive, and move on, and you free up your capacity to shape life anew in each moment.

Cynthia, the daughter of my old school friend, Faye, had not spoken to her mother for more than fifteen years. As a child, Cynthia had been sexually abused by a neighbor. Though she was sure that her parents did not know about the abuse, she nevertheless felt angry that her mother had failed to protect her. As soon as she graduated from high school, Cynthia left home and cut off all contact with her parents.

Faye could not understand what she had done to cause Cynthia to stay away. Though Cynthia did not answer Faye's letters, Faye took every opportunity to find Cynthia and maintained the hope that Cynthia would return someday. Though the pain of separation from her daughter was always with her, Faye refused to remain frozen in the past. She held long conversations with Cynthia in her mind, as she imagined her daughter would be today. She often talked about Cynthia to her friends so as to keep Cynthia present in her life.

Faye also took every opportunity to grow in her own life. Moved by the plight of Southeast Asian orphans, she and her husband, Don, decided to adopt a Cambodian child. The process of applying for the adoption was long and complicated. Faye and Don attended classes for prospective parents, acquired visas, got shots, assembled necessary documents, and made plans to travel to Cambodia to pick up their new daughter. Faye held the picture of the girl while she disembarked at a rural airfield surrounded by deep green hills. As she scanned the faces of the people meeting the plane, Faye caught sight of a tiny girl with sad, beautiful eyes and a incandescent smile. The answering smile in her heart told Faye that she had found her daughter.

Back home, Cynthia was making changes, too. Her excellent performance at the entry-level bank job she got after high school had led to several promotions. Eventually, Cynthia was offered a managerial position. She was finally feeling better about herself, which made her feel better about her mother, as well. Soon after Faye returned from Cambodia, Cynthia called her mother, and was welcomed home joyfully by her parents and her new sister.

The happy outcome for Faye and her family is directly tied to Faye's decision to find ways to expand her love as a mother rather than remain frozen in mourning over her lost daughter and Cynthia's willingness to call her mother after fifteen years of no communication. Faye's willingness to continue to love in spite of Cynthia's absence kept her growing and available to new possibilities. Rather than bear a grudge, Faye kept her heart open and thus was able to welcome both Cynthia and her newly adopted daughter into her life. Cynthia, feeling better about herself, was able to open to her mother's continued efforts to keep in contact with her.

Growing the Brain

Recent discoveries have generated a new understanding of the brain with profound and exciting implications. Researchers have discovered that new neurons are born in the adult brain. Such new growth does not happen in all regions of the brain, nor are as many neurons born in adulthood as there are during gestation and childhood. Nonetheless, the long-held belief that the physical structure of the adult brain is static, rather than plastic, has now been discarded.

Every new neuron that survives gives us enhanced capabilities for expanded brain functioning, via the new pathways formed. Research also indicates that sustained stress and the secretion of glucocorticoids (in chapter 2, we looked at the effects of cortisol, secreted by the adrenal glands in response to stress) reduces the number of neurons born.[6] Is it any wonder, then, that we are less creative when we are under sustained stress?

]OOO[

GROWTH, IT SEEMS CLEAR, is essential to enhancing our quality of life. Exploring new experiences stabilizes new neurons, actually growing our brains. Learn, explore, grow! This is the message that our brain sings out to us.

Growth occurs when we bring change into our lives. The changes can be small ones—finding a new way to drive to work or to a friend's house, learning to use a new software program, taking a course in conversational Italian. Big changes, such as Faye's adoption of her Cambodian daughter, don't happen every day, yet they are an enormous trigger for growth. Making changes helps us "birth" our lives moment-to-moment through adulthood and into a vibrant and active old age. "Birthing" our lives refers to living our lives writ large from the fullness of who we are.

BIRTHING YOUR LIFE

Find a quiet place to center yourself. Take a deep breath. As you exhale, say "Ahhhhhhhhh!" Do this several times until you have released whatever tension you are holding in your body and are deeply relaxed. As you read each question below, inhale again deeply. Allow the question to linger in your mind and percolate through your body. Repeat the question, slowly. Be mindful of the feelings that emerge. Take time with each question and allow the answer to emerge from deep inside.

[] What have you been yearning for in your life?

How have you nourished this yearning this week, this month, this year? Imagine that whatever you yearn for is present in your life now. Catch a glimpse of how your life would be if this yearning were fulfilled. Picture the new situation as clearly as you can. Allow the emotions connected with this new situation to rise. Stay with the experience and allow it to unfold.

[] What changes would you need to make to bring what you are yearning for into your life?

[] What is the first small change you could make?

[] What would be the consequence of this change?

If anxious feelings arise—or if you have the thought, I can't do that!—don't lose your focus. Instead, ask yourself:

[] What fixed beliefs that I hold prevent me from making this small change?

Breathe deeply. Review the thoughts that emerged when you felt you could not make the change. Did you think: I've never done that before. What makes me think I can do it now? Examine carefully the thoughts and beliefs that stop you. Now ask yourself:

[] What do I need to do to trigger this change?

Think of what you need to support you in deciding to make this change. Is it a willingness to let go of old pains and move in a new direction? Is it the conviction that you have suffered enough? Is it the desire to reclaim your life?

For some of us, the barrier to new growth is an old resentment or hurt from the past. Allow your mind to explore this possibility. Ask yourself:

[] Do I need to forgive someone who hurt me deeply?

If the answer to this question is yes, imagine that you are in a prison, kept there not by the person who hurt you, but by your own unwillingness to forgive and release the past. Imagine that you hold the key to unlock this prison door. Ask yourself:

[] Am I ready to forgive, to let go of the past and move on?

Now, allow yourself to consider what external steps you are willing to take to support your decision to make a change. For instance, ask yourself:

[] What kinds of new learning might support my decision to change?

[] What resources can I tap to provide this new learning?

Allow your mind to explore the many resources available to you, such as 12-step groups, seminars and workshops, Internet communities, church and meditation groups, therapists and personal coaches, and artists' and writers' circles. Allow your mind to range over the many educational and personal growth resources in your area, including extension courses at local colleges and universities, park and school district programs, recreational and health club classes, neighborhood and community service groups. Consider how any of these opportunities might help you generate new ideas, explore new directions, make new friends, and add new patterns of behavior to support the personal changes you wish to make.

Remind yourself:

I am responsible for birthing my life.

Only I can make the change I yearn for.

]0000[

CELLULAR WISDOM ENCOURAGES us to expand our capacities. It provides new neurons to assist us in learning things we have never learned before, in doing things we have never done before. We have the power to change our beliefs. By committing ourselves to learning and growth, we can make the small and big changes necessary to birth our life.

CHAPTER NOTES

1. Elizabeth Gould et al., "Neurogenesis in Adulthood: A Possible Role in Learning," *Trends in Cognitive Sciences* 3, no. 5 (1999): 186–192.

2. Seiji Miyata and Glenn I. Hatton, "Activity-Related, Dynamic Neuron-Glial Interactions in the Hypothalamo-Neurohypophysial System," *Microscopy Research and Technique* 56, no. 2 (2002): 143–157.

3. William Beecher Scoville and Brenda Milner, "Loss of Recent Memory After Bilateral Hippocampal Lesions," *Journal of Neurology, Neurosurgery & Psychiatry* 20 (1957): 11–21.

4. Suzanne Corkin et al., "H. M.'s Medial Temporal Lobe Lesion: Findings From Magnetic Resonance Imaging," *Journal of Neuroscience* 17, no. 10 (1997): 3964–3979.

5. Menahem Segal, "Rapid Plasticity of Dendritic Spine: Hints to Possible Functions?," *Progress in Neurobiology* 63, no. 1 (2001): 61–70.

6. Nicholas B. Hastings, Patima Tanapat, and Elizabeth Gould, "Neurogenesis in the Adult Mammalian Brain," *Clinical Neuroscience Research* 1, no. 3 (2001): 175–182.

CHAPTER 4

actualize life's potentials

THE DEVELOPMENT OF THE HUMAN BODY is a wondrous unfolding of potentials, the actualization of the miracle of life. The human body, with all of its organs and systems and its trillions upon trillions of cells, develops from a single cell, the fertilized egg. Yet a snapshot of the developing fetus taken at an early stage in this process might surprise you. The being you would see, with its tail and gills, might look to you more like a monkey or fish than a human fetus. Only after the process is complete can you appreciate how the myriad components and processes have unfolded seamlessly to create the finished product—a human being.

Actually, all of life is a process of unfolding and activating unseen potentials. While the process is underway, we often cannot foresee how things will turn out. This truth applies to many aspects of our lives. This chapter explores four related aspects of the process of development. By looking more closely at fetal development, we'll try to gain insight into how we might actualize our undeveloped potential.

First, we'll look at critical periods in the development of the fetus, times of particularly heightened sensitivity and converging influences. All growth, including foundations for future development, happens as a result of converging influences present during such critical periods. Next, we'll look at how cells influence each other during the development process. Within the developing fetus, cell-cell interactions provide cues that guide cells to migrate and differentiate or become specialized. Then, we'll explore the role timing plays in the development process. As the fetus matures, cellular events occur in sequence, one step after another. The sequence cannot be altered; each stage activates new potentials and sets up conditions necessary for the ensuing step. Finally, we'll examine the role of stem cells, from which all other cells arise. We'll discover that, though development does lead to stable structures and functions, some aspects of the development process are ongoing throughout life.

Critical Periods

The spinal cord is the simplest component of the central nervous system. Essentially a tube, the spinal cord contains sensory neurons that face the back of the body and motor neurons that face the front. Sensory neurons transmit messages, such as feelings of pain and pleasure, from the outer world to the brain; sensations of spatial position and temperature; and other information about the world outside the body. Motor neurons, on the other hand, transmit messages from the brain to the muscles to control their contractions. It is the contraction of muscles, of course, that allows us to pick up a pen, walk, play the piano, and perform every other kind of physical movement.

How do sensory neurons and motor neurons developing from the single cell from which we each began find their respective places in the spinal cord? In general terms, at a critical period in the development of the fetus, multiple influences converge to induce particular cells to generate internal signals that cause them to develop into motor neurons or sensory neurons.

Let's look at a specific example of one influence essential to the development of motor neurons. For a particular, critical period in the development of a fetus, a strip of non-neural cells at the edge of the developing spinal cord, secrete a protein that researchers have whimsically named Sonic Hedgehog. Once it has been secreted, Sonic Hedgehog diffuses from this strip of cells into the interior of the spinal cord, creating a gradient. The situation is much like a sugar cube dissolving in a cup of tea. More of the sugar molecules are located near the sugar cube than, say, at the surface of the tea. In the same way, more Sonic Hedgehog is present near the cells that produced it, then near the center of the spinal cord, with a gradient of progressively lower concentrations of Sonic Hedgehog between the two places.[1]

This gradient controls the expression of proteins by other cells. Developing motor neurons closer to the cells where there is a greater concentration of Sonic Hedgehog, are directed by its signal to produce more of a special protein (we'll call it transcription factor 1) and less of another special protein (transcription factor 2). Cells further away, where there is less Sonic Hedgehog, receive a different message, and so produce different concentrations of the two transcription factors. Along the gradient from higher to lower concentrations of Sonic Hedgehog, various combinations of the two transcription factors function as a switch to determine the developmental course for particular cells. As some transcription factors within developing cells are turned on and others are turned off, those that will become motor neurons are forever differentiated from those that will become sensory neurons. Once the future form and function of a cell has been determined, its future is committed.

What happens next is that developing motor neurons begin to organize themselves into pools of cells, forming networks that control distinct muscle groups. The influences of many, many transcription factors induced or repressed by signaling molecules, of which Sonic Hedgehog is only one, combine during this period. Circuits that control the muscles of the face, arms, fingers, torso, legs, and feet are thus differentiated. The convergence of these multiple influences produces the fundamental organizational structure of the spinal cord. The window of time in which all this complicated signaling activity occurs is called the critical period of this phase of fetal development.

]OOO[

WE FACE MANY SIMILAR CRITICAL PERIODS of transition and opportunity in our lives. In times of ambiguity, when our future course is unclear, our situation is not unlike that of undifferentiated cells. We, too, carry within us unexpressed potentials, genes of future possibilities that await the right combination of signals to begin to express. When life's circumstances combine to put us into such an undefined condition of being, we often experience anxiety and a sense of urgency to bring this unsettling feeling to closure. Ambiguity is difficult to sustain.

However, as is the case with developing neurons, multiple potentials exist only until all the components of the differentiation program—the composite of elements operating in a particular sequence that will induce the change—are in place. Once defined, the program unfolds in a specific direction. In the developing spinal cord, a combination of signaling molecules from outside and transcription factors expressed from within activate the potential of a cell to become a specific kind of neuron with a functional role in a specific circuit. During a critical period in life, the signals we allow to influence us and the internal choices we make in response to those signals often set a series of events in motion that commit us to a particular future.

Perhaps the lesson we can take from this aspect of cell development is: *Slow down!* During a critical period, we are too sensitive and too vulnerable to make quick choices. Rather, we should hang out in the flow for a while, exploring various perspectives and allowing new influences to converge. Paying attention to how we feel as we try out various choices will give us important information. Once we've activated a program of differentiation, it's often difficult to go back and start over.

Divorce, serious illness, the death of a loved one, losing a job, bankruptcy, or deciding whether to have a child or start a new business are just some of the critical periods we may face. If we move too quickly at such times, before the multiple influences that can help us choose our future course have had time to converge, we may make decisions that do not reflect what we want to express in our lives.

For instance, if we take the first job that comes along after being laid off, or remarry too quickly after a divorce, we have not allowed for other possibilities to percolate through us like the host of signaling molecules and transcription factors in the developing spinal cord. But if we move slowly and with awareness, our periods of transition can become opportunities to transform our lives toward fuller self-expression and greater wholeness.

Cecile took advantage of an opportunity for positive transformation when a series of events created a critical period in her life. When she married Howard, Cecile had no idea that she would live in fear. Over the years, the frequency and intensity of Howard's abuse escalated from an occasional slap to full-scale beatings. After each beating, Howard was always sorry and promised he would change. The last time he abused her, she had to go the emergency room. Cecile lied to the doctor, telling him that she had tripped on the stairs.

When she came home from the hospital, Howard even cried and bought her flowers. But Cecile had come home with more on her mind than her broken wrist. Tests done while she was in the hospital had confirmed what she suspected: she was pregnant. Cecile was terrified at how Howard might react to this news. She had kept his behavior a secret for so many years, but that would be more difficult now with a baby coming.

The next day, a letter from her sister Josie arrived for Cecile. Josie wrote that she and her husband, Tom, were coming through town the next week and planned to visit for a few days. Here was something new for Cecile to worry about. Should she tell Josie about the baby? Would Josie notice how Howard's mood soured as the evening wore on? Would Howard drink like he often did? Would he insult Josie and Tom? Cecile didn't know if she could stand the worry and the anticipation of finding out what would happen.

As the days passed, Cecile began to imagine that the child she carried was already someone she could talk to. She spoke softly to the baby, her hands pressed to her belly, promising that she would do whatever was needed to keep it safe. In the evening, she was quiet and withdrawn. Howard began to notice. "What's wrong with you?" he asked over and over, growing more and more impatient. Cecile knew she could not stall forever. Besides, Josie was coming in the day after tomorrow.

Cecile began to see that she could not raise a child in that house, living with Howard. She wished more than ever that she had someone to talk to. But who? She had no close friends; she had been afraid to invite people into the house. Her parents had been killed in an auto accident several years before. She didn't belong to a church or have a regular family doctor. There was no one . . . except, maybe, Josie.

But if she told Josie about Howard and the baby, what difference would that make? Telling someone wasn't enough. It wouldn't help her protect her child. Slowly, an idea began to take shape. She would have to do something drastic. She would have to leave Howard! Several times over the course of her marriage, Cecile had thought of leaving. But she had not known where to go. Besides, she had told herself, if Howard suspected she wanted to leave, he'd beat her even more. But Josie's visit and the baby coming helped Cecile push through her doubts and form a plan. Maybe, just maybe, she could leave with Josie and Tom. Once back in Chicago with them, she'd figure out what to do next. As she thought about this possibility, her heart raced and an unfamiliar feeling of hope rose inside her.

But what could Cecile ask Josie to do? She and Tom were not wealthy and their apartment was too small for her and the baby to live there. When doubts like these arose, Cecile shut them off quickly. I just have to do what I can do, she told herself. Leaving with Josie and Tom is a first step. It is my chance to get out of here, before Howard hurts me or the baby. Howard likes Josie and Tom. He said they could visit. When they leave, I'm going, too. I've decided. This is the time for me to make my move.

The first step in activating new potentials is to recognize when you are entering a critical period. Sometimes the signals are obvious, such as the onset of an illness or the loss of a job. Josie's pregnancy, for example, changed how she viewed staying in her marriage. Other signals may not be so apparent. You may feel uncomfortable or distressed for a long while without being sure about the source of your emotion. If the opening stages of a critical period go unrecognized, you may miss an opportunity to make important changes. But when you cultivate the awareness that a period of transition is upon you, you have a better chance of using the multiple resources available to help you activate new potentials. Here is a process that can help.

RECOGNIZING A CRITICAL PERIOD

Sit in a quiet place where you will not be disturbed. If you are agitated, add an element to your environment to help you relax. Soft music, candles, or flowers can help you create a peaceful, calm environment, a space where you feel safe and supported. Take a deep breath and release your tensions on the out breath. Do this a few times until you feel relaxed.

Consider the following questions and allow feelings and images to arise effortlessly. Simply allow them to move through your mind and your feelings. At the end of the session, you may want to record your responses.

[] Have you been feeling increasingly distressed or simply at a loss about your life lately? If so, allow the distressed feelings or the confusion to be present now. Sit with these feelings and thoughts for a while, knowing you are in a safe environment.

[] Now ask yourself: When did you first start to have these feelings and thoughts?

[] What major events took place in your life about the same time that these feelings and thoughts emerged?

[] What other times in your life did you experience similar feelings and thoughts? How does that time compare with this period? What did you do then?

[] How do you feel right now? Allow an image or mental picture to emerge that captures that feeling. If you wish, draw this image and label its parts or feelings.

If the answers to these questions indicate that you are entering a critical period, relax! Remind yourself that times of change in your life are simultaneously scary and exciting. During such times, converging influences create a window of opportunity for the unfolding of new life. Previously untapped potentials in you are waiting to be born.

Be aware of any anxiety you feel and avoid the tendency to escape ambiguity by settling for a quick fix of certainty. Resist the influences of society or family to make a decision or get on with your life. Don't short-circuit the process of transformation. Superficial change is not what critical periods are about. True transformation requires that you allow authentic change to emerge from your depths. Remember, your life unfolds from the inside out. Be patient and alert to what is going on inside. Engage in activities that give you the energy to sustain this period of transformation.

Take comfort in the fact that critical periods are, by definition, limited. They do not last forever. If you give the process space and time, it will come to completion. You will recognize signals that your transition time is ending. A shift in perspective will demand your attention. Suddenly, you'll see things you didn't see before, or you'll see things very differently. Allow your process to reach a natural resolution. And, all the while, be mindful of the calls of your heart.

Cell-Cell Interactions

We have been describing fetal development in terms of separate processes. However, many processes occur simultaneously during development. Discussing them separately helps us focus on the unique characteristics of each set of events, but we should keep in mind that, just as in our lives, many events happen at the same time, and they often influence each other greatly. Such is certainly the case with the cell-cell interactions that help to form the human brain.

Where a neuron is located within the brain determines what it does. Identical neurons located at different parts of the brain function in different ways because conditions at each region of the brain are unique. Moreover, during development, neurons born in another part of the nervous system must migrate to their appropriate places in the brain.

We can surmise how this amazing migration might occur by looking at how cells reach their ultimate position in one part of the brain, the cerebral cortex. The cerebral cortex is the outermost layer of the largest part of the brain, the cerebrum. The human cerebrum is divided into two halves, or hemispheres, and is made up of six layers of neural cells. The cerebral cortex is the area of highest functioning within the nervous system. It directs, coordinates, and mobilizes all other components of the nervous system to allow us to move, think, speak, see, hear, and make sense of the world.

The neurons that make up the layers of the cerebral cortex are born at different times during development. Younger neurons stack on top of older neurons. The newest neurons to form must travel through layers containing the oldest neurons before they take positions in the cerebral cortex, far away from the site of their birth.[2] Non-neural cells form a scaffolding along which these neurons travel. Any disruption in this pattern of migration can lead to serious problems in brain function, including mental retardation. Thus, cell-cell contact—some cells supporting and motivating other cells—is essential to the formation of a healthy brain.

]0000[

LIKE DEVELOPING NEURONS, we often need outside help to move us toward the highest expression of our life purpose. The lesson we can draw from migrating neurons in the developing brain is to look for the signals around us, in the vicinity of our current life. That which attracts us, pulls us, seduces us, may be our call to self-expression.

Sometimes a hobby or outside interest can provide the spark that motivates us to make fundamental changes. Remember Martin, the frustrated biomedical researcher we met in chapter 1? Photography was the one bright spot in Martin's work life. It was the spark of energy that unfolded a new path for Martin and allowed him to find his proper place in the world—the position from which he could function in the most effective and joyous way.

Other times, the spark is a new friend or relationship, or a new place to which we travel or move, that provides the cell-cell interaction that gets us moving. Even if we are thoroughly dissatisfied with life, somewhere in our vicinity is a spark of energy. To actualize life's potentials, we must let everything else drop away and look for that spark. Like the surface molecules that guide the migrating neurons, that spark can be the guiding light that leads us to where we belong!

FINDING THE GUIDING SPARK

In a quiet place, relax and create an atmosphere of comfort. Then take up your journal and record your answers to the following:

[] List the hobbies, activities, relationships, and places to which you have been most powerfully drawn in your life. Take the time to look back over old photographs or reread passages in your old journals to get in touch with those things that provided a guiding spark at some point in the past. Take the time to allow the memories and feelings that surround these things to arise and move through you.

[] Run your mind over your current life and list those things that generate similar feelings now. Be a detective. Assume that everything that comes to mind may be an important clue. If feelings of pain or frustration arise as you scan your life, gently move your mind toward more positive emotions. You are on the lookout for energizing influences—guiding sparks that can help you activate your highest potential.

[] Take a few moments to write about what you've learned about yourself. What kinds of activities, people, or places pull you? How might you bring more of these experiences into your life? What do these outside signals tell you about the highest expression of your life purpose?

Sequence

During the development of the nervous system, many events occur in carefully programmed sequences. The nervous system begins as a fold in a flat sheet of cells. This fold then moves and closes to become a simple neural tube. Through a step-by-step process of cell growth and differentiation, this tube develops into the entire nervous system, including the brain. The parts of the tube that will become the spinal cord change only a little. Other parts expand, bend, or crinkle to form the largest part of the brain, the two hemispheres of the cerebrum.

During this process of transformation, many genes are activated in specific order. For example, signaling genes initiate and coordinate cell differentiation. Those at the outer rim of the cerebrum delineate the boundaries of areas of the cortex.[3] If one of them fails to activate, a part of the brain does not develop.[4]

Sequence is also important to the development of motor neurons in the spinal cord. We've already talked about how the protein Sonic Hedgehog, secreted by specialized cells, controls the development of motor neurons. But what makes the specialized cells secrete Sonic Hedgehog in the first place? A structure called the notochord, which exists for only a short period of time within the developing fetus, sets this critical process in motion.[5] Once it has secreted Sonic Hedgehog, the notochord becomes part of the spinal vertebrae or backbone. Thus, the nervous system unfolds in a sophisticated sequence of carefully choreographed steps. Each step takes place at a particular stage in the process, and the order of steps cannot be altered or reversed.

]OOO[

ALTHOUGH TIMING IS OFTEN LESS CRITICAL when we're making changes in our lives, we can identify a number of distinct stages in the process of personal transformation. When we enter a critical period, for example, we are often confused about what to do or what direction to take. In chapter 1, we met Victoria, whose life changed significantly once her children left home. The confusion Victoria experienced was about her role. Because she was no longer an active mother caring for her children, who was she? Questioning our identity is a recognizable stage in most people's process of personal transformation, as I discovered through my own experience.

I could not have anticipated how profoundly my decision to leave academic life would affect me. I had made many changes in my life. I had entered a convent and left it. I had married and divorced two husbands before marrying a third time. Throughout these changes, however, I maintained one consistent element—my academic career. Even my entering the Dominican order was motivated by the goal of becoming a college teacher.

One Saturday morning, just after I had made the decision to leave academia, I fell into what I can only describe as a black hole. I recall asking my husband to leave me alone for several hours. Lying on the couch in our living room, I felt completely torn away from everything I knew myself to be.

When I closed my eyes, I had the sensation of falling through black, empty space. There was nothing to grab hold of and no bottom to the hole. It was profoundly disturbing to fall and fall, yet I knew that I should not terminate the process prematurely. I allowed this lost feeling to continue for several hours. Afterward, I experienced a sense of completeness I could not interpret. I shut down my analyzing mind and allowed myself to stay in that sense of peace.

Leading up to this event, the steps in my process of transformation were more ordinary. Feeling restless as I wrestled with my decision to leave academic life, I had hired a coach to help me figure out what I wanted to be and to do. At the time, I thought the coach had helped me find some answers. Yet my black hole experience took my loss of identity to a deeper place, one where I had absolutely no idea who I was.

For the six months after my black hole experience, I explored my internal landscape deeply. I repeatedly asked myself: Who am I? Who is the woman underneath the roles I have played? Quietly, I listened. I began to feel, rather than think, my way. I wanted the freedom to express my creativity without restraint, to flow from the inside out. I journaled constantly during this process. Feelings I had buried for many years surfaced. For instance, my grief at my father's death when I was ten years old emerged fresh and potent. I cried as I read the description of his funeral to a writing group I was working with in Hawaii. In giving myself time to excavate the layers of my psyche, I was coming to know my whole self, not just my intellectual side.

As my process continued to unfold, I began to experience important insights. I realized, for instance, that the aspect of my academic career that brought me the most joy and fulfillment was seeing the medical students I taught begin to understand the human nervous system. I thrilled at every "aha!" breakthrough in comprehension. Those moments were my sparks of energy. Looking further back, I realized that I had enjoyed teaching chemistry and math at Dominican College for the same reasons. The smile that crossed a woman's face when she realized that she understood a complex mathematical concept delighted me. I began to see that, for me, teaching was a way of serving others, of helping them develop their full potential.

Now I looked for ways to use this insight to birth a new career. The profession of coaching was getting to be better known at this time. I realized that, at its heart, coaching was a kind of teaching. So, I contacted the coach I had worked with and asked if she would mentor me as I began coaching clients in academia. This step established the trajectory leading to the professional life I enjoy today. Though this step-by-step unfolding is apparent to me now, it was not so obvious while I was in the midst of my transformational process.

Building on what I've learned through my own experience, I try to help my coaching clients recognize the stages of their own process. Their experiences confirm that personal transformation often unfolds sequentially with seven steps:

1. Restlessness, discontent, or dissatisfaction with some aspect of your life.

2. A triggering event that leads to separation from that aspect.

3. A period of feeling lost or confused about your identity.

4. Time spent exploring and reconnecting with yourself, with the aim of finding out who you are underneath the roles you play.

5. The recognition of sparks of energy and threads of joy and fulfillment in your life.

6. Exploration of how these guiding sparks might lead you in a new direction, one that utilizes your past experience and is consistent with what gives you joy.

7. An initial step that establishes the trajectory for the next stage of your process.

Take a few minutes to ask yourself where you are now in this sequence in your own process of personal transformation.

Stem Cells—Portals to Power

Stem cells have more potential for transformation than any other cells in the developing fetus. They can divide without limit and become other specialized kinds of cells, including neurons, red blood cells, muscle cells, liver cells, and bone cells.

As a fetus develops, its stem cells differentiate into every other cell type in the body. Researchers are currently trying to understand what controls the transformation of stem cells into various kinds of specialized cells.[6] They are asking what cellular events cause the genes that determine a cell's future identity and function to turn on or turn off so that the cell becomes, say, a bone cell rather than a liver cell.

Such research is fueled by the phenomenal potential of stem cells for therapeutic use. Imagine if stem cells could be used to grow new heart cells to repair tissue damaged by a heart attack, or replace the damaged neurons that cause Parkinson's disease, or replace pancreatic cells that do not produce insulin and cause diabetes.

Researchers used to think that stem cells of a particular tissue, such as those that replace red blood cells in the body, could generate only cells of that type. But recent studies using animal tissue have shown that bone stem cells can be induced to produce liver cells, muscle cells, neurons, and non-neural cells.[7]

Another astounding breakthrough is the discovery that stem cells exist in the adult nervous system.[8] Those new neurons, birthed in the adult brain that we spoke of in chapter 3, arise from such stem cells. Research is beginning to suggest that adult stem cells can be stimulated by appropriate signals, such as growth factors, to give rise to many other kinds of cells. Continued medical research means that we can look forward to the day when a damaged heart or spinal cord can be regrown using the amazing potential of our own stem cells.

]0000[

HIDDEN SOURCES OF POWER, such as the potential typified by stem cells, are present in our lives, as well as our bodies. The first step in accessing this power is for us to realize that it is available. Many studies in educational settings have shown that what we believe to be true about ourselves can be either a portal to power or a limiting barrier. Children who are told that they are gifted usually perform exceptionally. Children who are told they are slow usually demonstrate little academic progress.

We acquired many of the beliefs that we hold about ourselves when we were very young. In a way, the situation is parallel to the importance of stem cells in the development of the fetus. Just as our adult body's tissues and organs were formed from the stem cells of the embryo, so, too, may we live 80 to 100 years with beliefs about ourselves that we formed within the first 5 to 10 years of life. However, the discovery that stem cells are present in adult tissues can encourage us to reexamine beliefs we hold that may be limiting us. The amazing potential for stem cells to repair hearts, kidneys, and other organs demonstrates that we do have the potential to generate a new belief system that can lead to a more vital, satisfying, and healthy life.

How many times have you been surprised at the change in behavior of the surviving partner when a wife or a husband dies? My father ran a grocery store. Before he died, my mother, who was a teacher, had never written a check. After his death, my mother had to learn how to do what was necessary to keep the business going. At first, she thought she could never master the many different aspects my father had incorporated into the store, including providing freshly cut meats and gasoline. But my father's brothers, all of whom had their own grocery stores, helped her learn what to do. Ultimately, she decided to sell the store, but she did manifest the hidden potential needed to run the business until it was sold.

FINDING YOUR POTENTIAL

In a quiet, reflective state, consider the following questions:

[] What potentials have you already activated in the process of developing your life?

[] What potentials would you like to activate in the next year? In the next five years? In the next ten?

[] In chapter 1 we talked about finding and activating your dreams. What potentials could help you activate your dreams?

[] In chapter 2 you took a stress inventory. What potentials would you like to develop to bring your life more into balance?

[] In chapter 3 you examined the changes you would like to make in your life, and the beliefs that limit you. What potentials could you activate to make changes to your belief system?

]OOO[

THE LESSON OF STEM CELLS IS CLEAR. The potential exists within us for amazing transformation. Use it, and we thrive. As we activate more and more of our hidden potentials, our wholeness increases and our lives develop toward fulfillment and joy.

CHAPTER NOTES

1. James Briscoe and Johan Ericson, "Specification of Neuronal Fates in the Ventral Neural Tube," *Current Opinion in Neurobiology* 11, no. 1 (2001): 43–49.

2. Pasko T. Rakic, "The Importance of Being Well Placed and Having the Right Connections," *Annals of the New York Academy of Sciences* 882 (1999): 90–106.

3. Clifton W. Ragsdale and Elizabeth A. Grove, "Patterning the Mammalian Cerebral Cortex," *Current Opinion in Neurobiology* 11, no. 1 (2001): 50–58.

4. Pierre Gressens, "Mechanisms and Disturbances of Neuronal Migration," *Pediatric Research* 48, no. 6 (2000): 725–730.

5. Margaret Hollyday, "Neurogenesis in the Vertebrate Neural Tube," *International Journal of Developmental Neuroscience* 19, no. 2 (2001): 161–173.

6. Sean J. Morrison, "Stem Cell Potential: Can Anything Make Anything?" *Current Biology* 11, no. 1 (2001): R7-R9.

7. See note 4 above; Dennis A. Steindler and David W. Pincus, "Stem Cells and Neuropoiesis in the Adult Human Brain," *The Lancet* 359, no. 9311 (2002): 1047–1054.

8. Arturo Alvarez-Buylla, José Manuel Garcia-Verdugo, and Anthony D. Tramontin, "A Unified Hypothesis on the Lineage of Neural Stem Cells," *Nature Reviews Neuroscience* 2, no. 4 (2001): 287–293.

CHAPTER 5

tap into renewal

SERIOUS DYSFUNCTIONS CAN SPIN OUR LIVES OUT OF CONTROL. Some of us may be addicted to alcohol or drugs. Others may be filling an empty place inside with an uncontrollable drive for money or success, or with a dysfunctional relationship to food, gambling, risk taking, sex, or even another person or life role, such as caring for an aged or infirm parent or engaging in a series of troubled love relationships. Sometimes, as we fulfill a particular role, our dysfunction gradually increases. Eventually, often as a result of some crisis or catastrophe, we realize that we have become a person we do not especially like. We may need to take drastic measures to bring our lives back into balance.

People with severe epilepsy could be described as neurologically out of control. We know much more about this condition than we used to. Greek physicians called epilepsy the falling down sickness.[1] But those of us who fall down in aspects of our lives are not stupid or possessed by demons, evil spirits, or devils, as was believed about epileptics in the Middle Ages. Many highly intelligent and creative people, including Socrates, Julius Caesar, Joan of Arc, and Swedish chemist Alfred Nobel, suffered from epilepsy. The treatment of epilepsy today does not involve smearing the mouth of a collapsed epileptic with human blood, as Greek healers prescribed, or forcing them to drink the blood of wounded gladiators while spectators looked on, as was done in ancient Rome. Nor do we view epileptics with fear and suspicion, or ostracize them as outcasts, as happens even today in some parts of Africa.[2]

Today we understand that epilepsy is a brain dysfunction in which neurons fire repeatedly, without essential periods of rest and regeneration. (We discussed the importance of these quiet times, or refractory periods, in chapter 2.) Many cases of epilepsy can be treated with medication, though in some people, the dysfunction is so severe that seizures do not respond to drug therapy.[3] In such cases, surgery—even radical procedures such as removing a portion of the brain—can bring the condition under control. As we have discovered, the brain has the ability to recover its functionality, even when a portion has been surgically removed. The amazing ability of the brain to renew itself gives us hope that we, too, can renew our lives when we have overcome a dysfunctional pattern of behavior.

Addicted to Success

Sometimes a dysfunctional pattern of behavior disguises itself as an opportunity to achieve a longstanding goal. I think, in this regard, of a man I know whose name is Frank. The car dealership Frank worked for specialized in expensive foreign cars. For more than ten years, Frank had looked on with envy at the financial rewards, bonuses, and gifts available to a successful sales manager. He hungered for the chance to win these for himself. A year ago, he was given the opportunity. It wasn't the best time for Frank to take on a big new job. His son was only six months old, and his wife had had a difficult pregnancy. She needed Frank around to help out more, not less. But how could Frank pass up this chance? It might never come again.

So, over his wife's objections, Frank became the manager of six salesmen. The owner of the dealership told Frank that he expected him to lead the dealership out of the slump of the past six months. Frank was determined to succeed, whatever the cost. He began each day with a sales meeting at which he profusely praised anyone who had made a sale and berated those who did not. He rewarded successful salesmen by directing hot prospects to them. When other salesmen objected that this tactic never gave them a chance to succeed, Frank was unsympathetic. "If you want to sell, sell," he told them. "The only thing we tolerate here is success." Soon he began to punish those who didn't sell by assigning them to the worst shifts and docking their vacations.

Inevitably, morale at the dealership began to falter, and sales continued to lag. Frank couldn't understand it. He felt as though he were working as hard as he could, certainly harder than any of the lazy salesmen. Wasn't he in his office an hour before the dealership opened in the morning, and didn't he take the late shift every night? Things at home were not going any better. Frank's wife, Elaine, complained bitterly about his long hours. She was blue and cranky and was often asleep when Frank got home.

Frank began to feel increasingly angry, both at home and on the job. Why couldn't Elaine understand that he was working for her, too, so that she could be happy and enjoy a few luxuries? Come to think of it, however, he wasn't very happy himself. He used to pride himself on being a joker who could always make his coworkers laugh. But he didn't tell jokes anymore, and nobody laughed when he was around. Most days, he felt ignored and even ridiculed by his staff. The worst part was that sales were down again this quarter. He told himself that he'd have to try harder. If he kept pushing, he just knew he could succeed. Something would happen soon. It had to!

Frank didn't have long to wait. One night when he got home, the house was eerily quiet. Elaine's clothes were gone and so were the baby's things. Then he saw the note on his dresser. Elaine had gone to California to stay with her sister for a while, "to think things over." She wasn't sure if she'd be back.

Frank went into the bathroom to wash his face. When he caught a glance at himself in the mirror, he hardly recognized himself. He had big circles under his eyes, and his face looked drawn and angry. *What's happened to me?* Frank asked himself.

In the weeks that followed, Frank began to realize his life had gone out of control. He wasn't a very good sales manager, that was for sure, and now his wife and his son, the two most important people in the world to him, had left him. His sense of humor and enjoyment of life had disappeared, and he had turned into an unhappy, critical, driven man he scarcely recognized. No wonder his salesmen didn't like him and his wife had run away. He didn't like himself very much, either.

]OOO[

THE DYSFUNCTION IN FRANK'S LIFE is similar to the dysfunction of neurons in epilepsy. Within the last ten years, we have begun to understand why neurons fire repeatedly and cause the seizures characteristic of epilepsy. As we learned in chapter 2, impulses are transported across the membrane of a neuron by the movement of small charged particles, or ions. As sodium and potassium ions move into and out of the neuron, an electrical impulse travels across the membrane of the neuron, an "action potential." After an impulse has been conducted, a refractory or rest period occurs, during which sodium and potassium ions move back out or reenter the cell. Once this re-balancing has occurred, the neuron is ready to carry another impulse. In epilepsy, however, the molecular mechanism that regulates the openings in the cell membrane that allow ions to enter and exit does not function properly.

The openings, or channels, in the membrane are not simply holes. Rather, they resemble chimneys, hollow tubes with structural components that open and close, like the flue of a chimney. When a neuron is getting ready to fire, the sodium channels open and sodium ions enter the interior of the cell. At the same time, potassium channels open so that potassium ions can flow out. The change in the distribution of charge across the cell membrane caused by the opening and closing of the ion channels creates an action potential, a change in the electrical potential between the inside and outside of the membrane when the neuron is stimulated. After the impulse has been conducted, the process is reversed; the potassium ions reenter the neuron and sodium ions flow out, reestablishing the membrane's resting condition.

The opening and closing of these channels takes place with exquisite precision within millionths of a second. However, in epilepsy, the channels do not function properly. As a result, instead of the cell membrane being charged for the normal interval of ten to sixteen milliseconds, it remains charged for up to ten times longer.[4] This prolonged charging of the membrane causes the neuron to experience a sequence of action potentials. A neuron cannot sustain this level of activity without damage.

A channel is composed of three to four different molecules—we might think of them as the "bricks of the chimney"—each the product of a different gene. Several genes have already been identified as causes for the dysfunctions in the sodium, potassium and, most recently, the calcium ion channels in epileptics.[5] Medical researchers are certain that because the normal balance of outflow and inflow of ions has been disrupted, epileptic neurons become hyperexcitable and fire repeatedly for prolonged and damaging intervals.

]0000[

WHEN SOMETHING IN OUR LIFE IS OUT OF CONTROL, we are often in a similar hyper-excited condition. In this state, it's almost impossible for us to receive the messages others are sending, or to tell others how we are feeling. Like epileptic neurons, our channels of communication are not functioning properly. Elaine had tried to talk to Frank many times about the situation at home, but he was always too tired, too preoccupied with the problems at work to listen properly. Frank's staff had also tried to tell him about the difficulties caused by his management style, but there, too, Frank refused to discuss it. He was so convinced that his approach was right, that he ignored multiple warning signals. Most serious, Frank was no longer listening to the messages from his own body/mind. Having shut himself off from the normal input and outflow of information, Frank drove himself to remain in a hyperexcited state of activity that would inevitably cause damage.

Dysfunctional Networks

Many of us have seen the frightening symptoms of epilepsy: uncontrolled, violent movements affecting most of the body; loss of consciousness; a temporary cessation of breathing; and, following the seizure, confusion. Such widespread effects do not happen because of the hyperactivity of a single neuron or even a few neurons. Seizures occur because whole populations or networks of neurons become hyperexcited and discharge repeatedly at the same time.

How do these networks of hyperexcitable neurons arise? As we discussed in the previous chapter, during development, neurons migrate to their proper places in the brain. The failure of neurons to migrate to their appropriate sites may contribute to epilepsy by leading the formation of dysfunctional networks or connections between neurons.[6] Thus, in addition to being an abnormality of individual cells, epilepsy is also a communal dysfunction.

If epilepsy is not diagnosed promptly and its seizures are not controlled quickly by appropriate therapy, significant and even irreversible damage to the brain can result. Repetitive seizures activate hundreds of genes involved in the growth of neurons and the formation of new synapses or connections. These genes may cause new dysfunctional networks to form. Therefore, especially when the brain is still undergoing development, prompt diagnosis and treatment is critical.

]OOO[

RECOGNIZING QUICKLY WHEN SOMETHING is fundamentally wrong in our lives is as important as recognizing epilepsy. Unlike Frank, whose problems began after he was grown and married, Patrice's dysfunctional behavior began in high school. A bright and energetic seventeen-year-old, Patrice was a sports enthusiast who excelled in basketball and soccer. Then, in the middle of her senior year, her teammates noticed that her reaction time was slowing. They could depend on Patrice less and less often to make the right move, shoot the ball through the hoop, or make a well-placed kick. She began missing practice and was often late for her classes. She also looked pale and was losing weight.

Patrice thought she was hiding her drinking well. She'd even managed to keep her secret from Ann, her best friend. The problem had started when she met Dan at a party in her junior year. He hung around with an older crowd that all drank heavily. At first, Patrice hated the taste of beer, but soon, she began to enjoy how giddy and free she felt when she'd had a few drinks. She didn't blame Dan and his friends; after all, it was her idea to hide a bottle at home so that she could have a drink or two to relax after school, and a few more before bed—to help her sleep, she told herself. She felt happier when she was drinking, as if the bad parts of the world had simply disappeared. Ever since that night when she had been woken up by her aunt and told about the car crash that killed her parents, she had wanted part of the world to disappear.

At first, Patrice loved the fun of sneaking around with Dan and his friends. Drinking on the sly became a game. But as the months passed, she wondered how long she'd be able to keep her drinking from Aunt Elsie and Uncle Fred, with whom she'd lived since her parents' death. The restaurant they owned took a lot of their time, and they were not around much, but Patrice was worried that someday they'd miss the bottles she'd taken from the restaurant bar, or find the ones she'd stashed in her closet. Could she stop? Did she want to? Patrice wasn't sure.

Patrice, like Frank, was in denial about the potential consequences of her dysfunctional behavior. Here is a questionnaire you can use to help you assess whether some aspect of your life is similarly out of control.

PERSONAL DYSFUNCTION QUESTIONNAIRE

Answer the following questions "yes" or "no." Record your answers for review later.

[] Do you often have occasion to apologize for a lapse in behavior?

[] Do you have frequent feelings of guilt, shame, or remorse?

[] Is there anything about your life you'd be ashamed to have people find out?

[] Are you consistently late?

[] Do you rely on some pattern of behavior to escape worries or anxiety?

[] Do you have difficulty sleeping?

[] Have you ever lost time from work or school because of over-indulgence?

[] Do you rely on some pattern of behavior to help you overcome shyness or to put you at ease?

[] Have people in your life expressed concern over some pattern in your behavior?

[] Have you ever experienced financial difficulties because of some pattern of behavior?

[] Have you ever considered consulting a physician for help overcoming some pattern of behavior?

[] Do you experience depression or anxiety if something prevents you from engaging in some pattern of behavior?

[] Are you preoccupied at inappropriate times by thoughts about a particular subject or pattern of behavior?

If you have answered more than half the questions in the affirmative, you may have a personal dysfunction, consider taking corrective action.

The Spread of Dysfunction

If epilepsy is not brought under control quickly, the condition spreads from one part of the brain to the other. What begins as a dysfunction of a small network of neurons quickly involves larger and larger networks.[7] Remember that a single neuron can contact hundreds, even thousands of other neurons. The initial site or focus of epileptic activity can occur in any part of the brain. If the condition begins in an area of the cerebral cortex that does not directly involve sensory or motor functions, it may go unrecognized until it spreads enough to damage adjacent areas that control these functions.

Epilepsy can even spread from one hemisphere of the brain to the other, using the commissures, or pathways, that connect the two halves of the brain. A different set of genes than was involved in the initial dysfunction may underlie the spread of epilepsy across networks. As more and more networks become involved, an epileptic person's ability to function normally is increasingly compromised, physically, psychologically, and socially. Uncontrolled epilepsy can also negatively impact learning. Thus, the condition requires immediate intervention to prevent further harm.

]0000[

OTHER TEENS WHO ENJOYED DRINKING were attracted to Patrice, Dan, and their friends. Patrice was amazed how many kids seemed to enjoy drinking as much as she did. When Dan's friend Mike helped her get a fake I.D. card so she could get into clubs, drinking and dancing became her life. She had loads of new friends to drink with and a new life that helped her forget the things she couldn't control. She met kids from every part of town and from many different schools. As the group of her acquaintances grew larger and became more diverse, Patrice felt safer. There was comfort in numbers. Everybody I know drinks as much as I do, she told herself. Except that she really didn't know much about her new friends. Only one thing united them: finding new places to drink together, and sharing stories about how they hid their addiction at home and at school.

There were times when Patrice wondered what would happen to her. She couldn't study the way she used to, and her grades were falling. The other day in Spanish class, she fell asleep when someone was talking to her. She used to love Spanish and was the one who spoke most often in class. Now she was falling asleep. How long could she keep this up? She didn't feel all that well, either; after a night at a club with her friends, her head ached and she felt shaky, unfocused, and anxious.

One day a friend of Dan's needed two hundred dollars to pay a court fine for underage drinking. When Patrice said she didn't have any money, Dan urged her to get some. Patrice was scared. She was already "borrowing" twenty-dollar bills from Aunt Elise's wallet whenever she could. Her aunt and uncle worked hard for their money, and they had given her a home. How could she steal from them? How much had Dan stolen, and what about the others? She'd heard that her new friend, Joe, had been suspended from school for taking money from a teacher's purse. Would they find out about her stealing and suspend her too? What would Aunt Elsie do if she discovered that Patrice was stealing and why she needed the money?

WHO OR WHAT SUPPORTS YOUR DYSFUNCTION

Here are some additional questions to help you assess whether you are surrounded by people or circumstances that evoke or support your personal dysfunction—write your answers for review later:

[] Do you notice that you engage in a potentially damaging pattern of behavior when you are with a particular group of people or in a particular circumstance?

[] Do you become more irritable with your family and others you are close to after you engage in this behavior?

[] Does your behavior minimize the time you spend with your family and others you are close to?

[] Have you ever committed or considered committing an illegal act to finance your behavior?

[] Do you seek out lower companions or an inferior environment when you are engaging in this behavior?

[] Do you wonder what will happen to you if you continue this behavior?

Intervention and Treatment

Many people with epilepsy can be treated with a combination of drugs. However, some 20 percent of cases are resistant or simply do not respond to drug therapy. The story of an epileptic girl told in the *Washington Post* illustrates that radical intervention and treatment can help even in resistant cases.[8]

Maranda Francisco's epilepsy was quite severe. By the age of four, she was experiencing up to 120 seizures a day. This level of hyperactivity was causing cell loss and other severe damage to Maranda's brain. Severe epilepsy is progressive. Repeated seizures lower the thresholds for future seizures, escalating their frequency and spreading the damage across more and more of the brain. Maranda's seizures could not be controlled by drugs. Data from cases like hers indicated that a spontaneous remission was very unlikely. Her doctors concluded that surgery was the only answer. Introduced in the late 1800s in the United Kingdom, the surgical removal of the affected part of the brain has been shown to improve or eliminate seizures in 70 to 90 percent of cases. Further, by removing the damaged neurons, surgery stops the debilitating effects of spreading dysfunction.

The procedure could hardly have been more radical: surgeons at Johns Hopkins University Hospital in Baltimore removed the left half of Maranda's brain! Though the left brain includes the centers that control language and speech, removing a smaller section of brain tissue would not have stopped the spread of Maranda's epilepsy. Without such drastic treatment, Maranda's entire brain would soon have been damaged.

]OOO[

A DIFFERENT KIND OF DRASTIC INTERVENTION was implemented in Patrice's case. Patrice feared she would be found out, but she knew it was inevitable. Her aunt had gone into Patrice's room to look for a pair of shoes she had borrowed. Patrice had grown careless and left a bottle under her bed, empty but still smelling of the vodka that was once in it. Finding the bottle confirmed a suspicion Aunt Elsie had been harboring for a while. She did not relish telling her husband, Fred, about her discovery, but she knew something had to be done.

Elsie and Fred arranged an intervention conference with Patrice's guidance counselor and a social worker for the school district. When she was confronted with the evidence of her addiction, Patrice broke down and began to sob. She told the group everything and became, once again, a scared little girl who knew she was in trouble and needed help. The social worker outlined various treatment options—local counseling, therapy, and support groups—but she felt that it would be difficult for Patrice to break off contact with Dan and his group if she remained in the area. When they heard how much Patrice had been drinking, Elsie and Fred decided that her problem was so severe that a more radical treatment was called for.

With Aunt Elsie and Uncle Fred's approval, the social worker arranged for Patrice to go to a residential treatment center for young alcoholics in the next state. Patrice was distraught and angry, but she was also relieved. The game she had been playing was over. Scared and trembling, she entered the treatment center.

Frank's marriage was beyond saving. During her time at her sister's, Elaine had come to realize that Frank's drive for money and prestige had always been more important to him than his family. Frank had grown up in a household where money was scarce. He had vowed never to be without. Over the years of their marriage, Frank's ambition had escalated until it was clear that he would always put his job ahead of his family. Elaine couldn't live that way, nor did she want her son to grow up under those conditions. She told Frank she was filing for divorce.

DETERMINING IF YOU ARE PREPARED TO CHANGE

The following questions can help you determine whether you are ready to make the radical changes that might be necessary to overcome a severe personal dysfunction:

[] Have you ever considered self-destructive behavior as a result of some negative pattern of behavior?

[] Have you ever promised yourself you would stop this behavior but been unable to do so?

[] If there is a circumstance or group of people that evoke or support this behavior, would you consider cutting them out of your life?

[] What other changes are you willing to make to stop this behavior?

]0000[

MAKING RADICAL LIFE CHANGES is certainly scary, but it is sometimes the only effective therapy. Changing jobs, ending a marriage or other partnership, moving to a new city, seeking help from a physician or therapist, joining a 12-step or other support group, even checking yourself in to a residential detox program can all be radical treatment options, depending on the type and severity of your dysfunction. Nevertheless, these changes, however drastic they may seem, are worth making if the alternative is lasting damage to your health or your life.

Growth and Renewal

Fourteen years after her brain surgery, Maranda was dancing, writing emails, and taking karate lessons. Although the part of the brain that normally controls speech was removed, Maranda was able to speak again.

How could this happen?

The body's capacity for renewal is truly astounding. Over time, the right half of Maranda's brain took over the functions of the left half, which had been removed. The process required extensive rehabilitation therapy and a lot of hard work and commitment, as Maranda relearned to talk, walk, interpret what she saw, and perform other body functions.

The brain's capacity for renewal is demonstrated in other cases, as well. Stroke victims and people who have suffered brain injuries often experience an amazing recovery of functions after trauma. Recent studies in people sixty to seventy years old demonstrate that aerobic exercises that increase the body's intake of oxygen, including such simple exercises as walking, enhance brain function—specifically a person's ability to perform executive tasks.[9] Other studies demonstrate that stroke patients can improve the use of their dysfunctional arm by looking at their functional arm in a mirror and attempting to imitate its movements using the dysfunctional limb.[10] Animal studies also demonstrate that exercise increases the number of blood vessels in specific regions of the brain.[11]

]0000[

PATRICE WAS WITHDRAWN AND DID NOT SPEAK UP in group therapy sessions when she first arrived at the treatment center. Within a few weeks, however, Patrice began to notice that the other people in the group had problems as devastating as hers. Each session, it was a little easier for her to talk. She began to ask questions of the others and to answer questions directed to her. When someone in the group asked her, "How do you feel when you drink?" she described how comfortable it was when the pain dropped away. Then the therapist who was leading the group asked, "What could you do to feel that comfortable without drinking?" Patrice had never asked herself that question before. It started a new train of thought in her that, for the first time in many months, made her smile.

Under the care and attention of the staff at the center, Patrice began to heal and to rediscover her ability to function without alcohol. Medications helped her deal with the physical aspects of her addiction and to overcome the need to have the world drop away. As she developed new habits and new perspectives, Patrice's fears slowly transformed into hope. Having her own children and creating a home for herself became easier and easier for Patrice to imagine as time went on.

Aunt Elise and Uncle Fred began to visit six weeks after she arrived. It was good to see them without worrying about hiding something. Patrice began to talk to them about her mother and father, which she had never done, and about her dreams for a family of her own. With the help of the tutoring program at the center, Patrice finished high school. She began to help with the other patients at the center, and organized teams to play volleyball and basketball. As she recovered her essential self, her future, once so gray, began to brighten and take shape.

Recovering Our Lives

Fear often stops us from making the radical decisions necessary to restructure our lives. The message of cellular wisdom—that the body, even the brain, is able to regenerate—tells us that we do have the resources within to recover our lives, even after serious trauma. Restructuring does not happen overnight, but it does occur, step by step, decision by decision. We can rebuild a life that supports us rather than one confines or destroys us. The following exercise can help you take the first step toward recovery.

TRANSCENDING DYSFUNCTION

Create a quiet place where you feel very comfortable, relaxed, and supported. You may want to keep in view photos of the people who love you and whom you love, to help create an atmosphere of loving support. Remind yourself that there is nothing to fear. Blame or recrimination is not part of this process. What's needed is truth, honesty, and rigorous self-assessment.

Take a deep breath, knowing that your breath brings nourishing oxygen to support your body's functions. As you release the out-breath, remind yourself that you're letting go of waste materials, things you no longer need in your life, limiting beliefs, toxic circumstances, and dysfunctional associates.

Take three or four deep and cleansing breaths, until you feel still, safe, and calm inside. Review the answers you gave to earlier questions in this chapter recorded for use here. You're going deeper now, confident that you will be able to change your behavior. You're ready to confront any dysfunction in your life. You realize this confrontation is the first, necessary step toward recovery. With great gentleness, allow your thoughts to explore the following questions and suggestions.

[] Is there any aspect of your life that is less functional than you would like it to be?

Recall, as clearly as you can, incidences when this less functional aspect is present.

[] How do you feel as you recall these instances?

[] When this less functional aspect is present in your life, is there someone consistently present?

[] What is the role of this person?

[] What might you do to alter your relationship to this person?

Imagine as clearly as you can the new relationship.

[] When this less functional aspect is present in your life, what circumstance or role is consistently involved?

[] How does this circumstance or role evoke your less functional aspect?

[] What might you do to alter the circumstance or role?

[] Imagine as clearly as you can the new circumstance or role.

[] If it is not possible for you to alter your relationship to the person or to change the circumstance or role, you may need to consider separating yourself from the these influences.

Imagine yourself doing this.

[] What are your fears? What are the limiting beliefs that underlie these fears?

[] Who might you call on for support during this process of separation?

Finally, imagine a scene in which your desire for change is fully manifest. How are you dressed? What are you doing? Who else is present? Imagine, as clearly as you can, your state of mind as you inhabit this mental picture of the functional self you wish to inhabit.

It is not necessary to plot out in advance a complete plan for change. Simply allowing yourself to get in touch with an aspect your life that is dysfunctional, and reminding yourself imaginatively that change is possible—so begins the process of recovery and regeneration. The desire for a fully functional life can seed whatever changes you need to make. Holding the desire can attract the circumstances that lead to restructuring and reclaiming your life.

CHAPTER NOTES

1. Jeffrey M. Jones, "'The Falling Sickness' in Literature," *Southern Medical Journal* 93, no. 12 (2000): 1169–1172.

2. Louise Jilek-Aall et al., "Psychosocial Study of Epilepsy in Africa," *Social Science & Medicine* 45, no. 5 (1997): 783–795.

3. O. Carter Snead III, "Surgical Treatment of Medically Refractory Epilepsy in Childhood," *Brain and Development* 23, no. 4 (2001): 199–207.

4. Ortrud K. Steinlein and Jeffrey L. Noebels, "Ion Channels and Epilepsy in Man and Mouse," *Current Opinion in Genetics & Development* 10, no. 3 (2000): 286–291.

5. See note 4 above.

6. Asuri N. Prasad, Chitra Prasad, and Carl E. Stafstrom, "Recent Advances in the Genetics of Epilepsy: Insights from Human and Animal Studies," *Epilepsia* 40, no. 10 (1999): 1329–1352.

7. See note 6 above.

8. Don Colburn, "Scientists Discover Brain's Adaptability," *Washington Post*, September 28, 1999, H12.

9. Arthur F. Kramer et al., "Ageing, Fitness and Neurocognitive Function," *Nature* 400, no. 6743 (1999): 418–419.

10. Eric Lewin Altschuler et al., "Rehabilitation of Hemiparesis after Stroke with a Mirror," *The Lancet* 353, no. 9169 (1999): 2035–2036.

11. Krystyna R. Isaacs et al., "Exercise and the Brain: Angiogenesis in the Adult Rat Cerebellum After Vigorous Physical Activity and Motor Skill Learning," *Journal of Cerebral Blood Flow and Metabolism: Official Journal of the International Society of Cerebral Blood Flow and Metabolism* 12, no. 1 (1992): 110–119.

CHAPTER 6

claim abundance

ABUNDANCE IS ALLURING. However much love, beauty, health, wealth, or pleasure we have in our lives, we always want more. We seem to be programmed with an innate drive to better ourselves, to improve on what we have, to be all that we can be. Our desire for material abundance is due, in part, to our extravagant culture. Advertisements entice us with promises of the pleasures of a new car, whiter teeth, the latest fashions, a fabulous vacation. Dreams of abundance induce sweet feelings of luxury and ease.

Not all experiences of abundance are sweet, however. Many of us have been stung by people who aggressively pursue the job that draws the highest salary, the most expensive home, the most lucrative investment portfolio, the most prestigious school. We may have experienced these out-of-control desires ourselves. As we witness the sometimes addictive grasping for more and more abundance, a question emerges: Can we learn to enjoy abundance gracefully, even to pursue it—without becoming enslaved by it?

Our body teaches us that balance is possible. Abundance, even extravagance, is evident everywhere in our physiological makeup. Yet mechanisms exist within our body systems to regulate and control abundance before it damages us. These mechanisms model ways for us to claim our abundance without losing our balance or our sense of proportion.

Extravagance and Letting Go

No dilemma surrounds extravagance in the body. The developing nervous system generates far more neurons than will be incorporated into the mature system. Similarly, millions of sperm are produced, even though only a single sperm is needed to fertilize an egg. Yet our body has the built-in wisdom to recognize that uncontrolled, unregulated growth can be harmful, even deadly. Cancer develops because the normal mechanisms that limit cell growth fail. In the healthy body, unneeded cells die. In this way, the body balances extravagance by letting go naturally of what is not needed.

As the nervous system develops, those neurons that will not contribute to mature functioning die. For example, as many as 80 percent of neurons that are generated to carry impulses from the retina to the brain—neurons critical to our ability to see—die during the process of gestation.[1] Neurons that make up the receiving component of the eye-brain circuit develop in the retina. As they grow, these neurons send their processes (the long tail that transmits the impulse) toward a target in a part of the brain called the thalamus. Only those neurons that are successful in connecting the retina to this target survive. The others die.

The death of these excess neurons is not induced by disease or trauma. The neurons are destroyed by the activation of "suicide genes" within the neurons. These genes seem to be turned on by a program from within the neurons, initiated by death-signaling molecules from the neurons themselves, or from surrounding tissues. At the same time, the developing brain target secretes substances that inhibit the activation of suicide genes in those cells that have successfully made the connection and become functional components of the eye-brain circuit.

Like excess neurons, sperm that do not succeed in fertilizing an egg also die. Each instance of ejaculation produces three hundred million to four hundred million sperm, clearly an instance of extravagant abundance. Yet the genetic makeup of reproductive cells makes them very fragile. A sperm cell contains only half the required number chromosomes. Female reproductive cells, called ova, also contain only half the required number of chromosomes. Neither kind of reproductive cell lives very long on its own. Only the fertilized egg, the union of the sperm and ova, has the complete set of chromosomes necessary to sustain life.

Perhaps the male body is so extravagant in producing sperm because the window of opportunity during which fertilization can occur is so limited, and the journey sperm must take toward the egg is so arduous. Sperm must swim upstream to reach the oviduct where fertilization takes place. They move at a rate of three millimeters per minute, so it takes a sperm thirty to sixty minutes to reach the egg. Only the most vigorous sperm get there; weaker sperm do not reach the egg in time. Abundance assures that some sperm are successful and that reproduction happens.

The message about abundance is clear: the body zealously pursues both generation and destruction, yet in a dynamic relationship, extravagant growth is balanced by death. Those cells that contribute to the appropriate functioning of the body are nurtured and supported; those that do not are let go.

]0000[

THE SAME BALANCE SHOULD BE TRUE IN OUR LIVES. Those aspects of abundance that support and nurture us should be fostered; those that do not should be let go. In a way, the process of shaping our lives so as to enjoy appropriate abundance is similar to the way Michelangelo is said to have sculpted his famous statues. He described his method as chipping away the excess marble to allow the figure of *David* to emerge from the marble block. We, too, can sculpt our lives by removing the excess and allowing our true shape to emerge. As a first step, we assess what is really important to us—what we wish to retain and what we wish to let go, what nurtures us and supports our true expression, and what blocks and inhibits it.

As long as we let go of that which does not advance us, we can pursue abundance with a passion equal to the body's extravagance. Holding on to what is extraneous—those things and experiences that do not advance us or contribute to our well-being and that of others—is terribly damaging. The key to thriving, joyful abundance lies in achieving a dynamic balance between generation and destruction, between extravagance and letting go.

SCULPTING A LIFE OF ABUNDANCE

Step One: Self-Assessment

Find a quiet place and settle in. When you are comfortable, breathe deeply. Slowly fill your lungs with oxygen and feel the life energy streaming into every cell of your body. Exhale and release everything that binds or restricts you. With the intention of allowing appropriate abundance to flow to you, allow your mind gently to scan your current life. Effortlessly, without tension, reflect on the following questions:

[] In what specific ways is my life rich and abundant right now?

[] How do these various aspects of abundance nourish me and support my true self-expression?

[] What single step could I take right now to increase my experience of appropriate abundance?

[] In what ways is excess present in my life right now?

[] How do these excesses block or inhibit my sense of well-being?

[] What single step could I take right now to let go of some aspect of excess, or diminish its presence or influence in my life?

After answering these questions, take some time to draw up a plan for increasing some aspect of appropriate abundance and for eliminating some aspect that seems excessive or that inhibits your well-being. If you wish, write your plan as a contract with yourself. Be as specific as possible, and make each step in your plan small and achievable. Sign and date your plan.

For instance, if you determine that good relationships with family members are an important aspect of your abundance, make a plan to organize a family reunion, or to send a weekly email to your sister in a far-away state, or to post photos from your vacation on a website to share with your family. If you decided that the clutter in your house is an aspect of excess you'd like to eliminate, schedule a day in the next week to fill boxes to give away to charity, or to plan a garage sale, or to decide what you might give away to friends as wonderful gifts.

Maximizing Potential

Genetics is big news these days. There was great excitement in the media when the team of scientists involved in the International Human Genome Sequencing Consortium announced that they had finished counting and mapping the genes in human chromosomes.[2] Yet the findings of the consortium also raised questions. Scientists had estimated that they would find one hundred thousand genes or even more, yet the first draft of the human genome map revealed that humans have fewer than half that number, approximately thirty to forty thousand genes. To put this number into perspective, human beings have about twice the number of genes as the Drosophila fruit fly, but about the same number found on the chromosomes of a corn plant!

Even if a few thousand additional genes are added to the human genome map, scientists must answer a difficult question: How can such a small number of genes generate the biological complexity of human beings? Only one explanation makes sense: Each gene must be utilized to its maximum potential.[3]

We know that human genes can generate the large number of proteins necessary to make up the complex organs and systems of the body by cutting and pasting together chemical components. This technique, known as alternative splicing, makes it possible for many different, but related proteins to arise from a single gene. Alternative splicing is most evident in the human nervous system.[4] More than five hundred alternative variants of one primary gene transcript are expressed differentially in the hair cells within the inner ear, giving us the capacity to hear a gradient of sound frequencies from high to low tones.

]OOO[

IT SEEMS CLEAR THAT BY CREATIVELY using their potential, human genes are able to accomplish the wonders of human complexity. The most creative people are also accomplished at maximizing their potential. Consider, for example, physicist Richard Feynman, whose contributions to quantum physics are unquestioned. In 1965, he was awarded the Nobel Prize for Physics. Author Michael Michalko reports that Feynman's IQ measured 122—certainly above average, but less than the range for people termed geniuses.[5] What accounts for Feynman's astonishing creative facility?

In his own books,[6] Feynman describes his joy in asking questions and his ability to turn a problem around and examine it from multiple angles. Because he loved physics, he continuously sought to challenge himself. Even the breakthroughs that led to his Nobel Prize, Feynman says, were the result of creative play: "The diagrams and the whole business that I got the Nobel Prize for came from that piddling around. . . ."[7]

Feynman's love for probing into how things work began in childhood. From the time he was eleven or twelve, he took radios apart and fixed them in a home lab. Innovative and creative, Feynman models for us a way to maximize our potential: Find something you love and explore the many different ways of expressing it.

A friend told me the story of Stephanie, the middle child in an accomplished Boston family. Her older brother, Joe, was a financial whiz who attended Harvard. Her younger sister, Lisa, a talented pianist, was headed for the Conservatory of Music. Stephanie's successes were not so celebrated. Often she felt as if she did not belong in her family. She wondered if her family even noticed she was there.

Stephanie's way of coping with being ignored was to create an imaginary family. In her fantasy, her more modest accomplishments were enthusiastically appreciated. Over the years, the characters within this imaginary family came to seem almost like real people. Her imaginary father read her school papers aloud and lavishly praised their insightfulness. Her imaginary mother was delighted by her creativity in gardening and flower arranging. Her imaginary older brother spent time teaching her about the stock market, while her imaginary sister always played new pieces for Stephanie first and prized her feedback.

In her last year of high school, Stephanie wrote a story for her English class that brought this imaginary family to life in a series of moving vignettes. Her teacher was so impressed with Stephanie's story that she entered it in a college scholarship contest. To Stephanie's surprise and delight, her story took first place. The prize was a year's tuition to study creative writing at a small New England college. Imagine Stephanie's pride when, after she broke the news to her real family, her father read her story aloud and praised its creativity and insights!

Like Stephanie, we can learn to maximize our potentials by leveraging what we have to achieve optimum results. The following exercise builds on the insights you gained from the work you did in manifesting hidden potentials at the end of chapter 4. Here, you explore ways to maximize your potential as a means of claiming appropriate abundance.

SCULPTING A LIFE OF ABUNDANCE

Step Two: Maximizing Your Potentials

Review your answers to the exercise for "Finding Your Potential" in chapter 4. Follow the method explained in Step One in this chapter to enter a place of quiet and relaxed reflection. Allow the energy streaming into your cells to help you to maximize your potentials, and reflect on the following questions and scenarios:

[] What potentials have you activated already in your life?

[] With great ease, imagine many ways you might use these potentials, in various circumstances, occupations, locales, and applications.

[] Which of these imagined ways seem effortless, bringing you joy as you visualize putting it into practice?

[] Now ask yourself, What potential might I activate to help bring the balance of appropriate abundance into my life?

[] In your mind's eye, imagine that activating this potential helps you to let go of some specific instance of excess and/or to increase your experience of abundance. Make your mental picture as vivid as you can. See the activation of this potential as effortless and as bringing you joy.

[] Now ask yourself, What potential might I activate to help me make necessary, but difficult changes in my life?

[] In your mind's eye, imagine that activating this new potential helps you make a specific and positive change. Make your mental picture as vivid as you can. See the activation of this potential as effortless and as bringing you joy.

[] Now ask yourself, What potential do I wish to maximize within the next year?

[] In your mind's eye, see yourself activating this new potential. Make your mental picture as vivid as you can. Imagine the reactions of people in your life, the effect your new activity has on your health, circumstances, and emotional outlook. See the activation of this potential as effortless and as bringing you joy.

[] If you wish, take some time to draw up a plan for activating this
potential as a contract with yourself. Be as specific as possible about
what you will do, but make each step small and achievable. Sign and
date your plan.

For instance, say that you have a facility for making desserts. You have
already manifested this potential in concocting wonderful treats for
your family and friends. Now allow yourself to imagine that you have
started a business making the cheesecakes everyone raves about for
local restaurants, or that you have begun giving classes in fancy
desserts at the gourmet cooking supply shop in your town, or that you
have written an article about your cakes that is published by a glossy
cooking magazine. Imagine the financial and emotional consequences
of activating this potential. If it seems appropriate, decide that as a first
step, you will fine-tune your five favorite cake recipes and choose one
to submit to a cooking magazine.

Complementarity and Redundancy

Abundance is also demonstrated in the body by the principles of complementarity and redundancy. *Complementarity* is the collaborative action of two or
more parts of the body that share a common characteristic. Any action carried out by complementary parts is more powerful than what either part
could do alone, as the shared action both reinforces what the two have in
common and uses their differences in service to a shared goal. *Redundancy*
refers to a characteristic shared by body parts. We might think of it as a kind
of insurance policy. If one part cannot carry out an action, another part with
the same characteristic can. Within our cells, as within our lives, complementarity and redundancy unlock new sources of energy and power.

The simplest example of complementarity and redundancy occurs within
a single cell. As we saw in chapter 1, the DNA in the nucleus of a cell holds
the blueprints to produce molecules within the cell. We learned in chapter 2
that another organelle within the cell, the mitochondria, is an energy factory
that drives the activity of molecular production. Like the nucleus, the mitochondria also contains DNA. Mitochondrial DNA contains the codes to
generate molecules that transfer energy within the cell. The complementary

link between the instructions contained in the nucleus and the energy to carry them out is DNA. Though the task of the DNA in the nucleus and the DNA in the mitochondria is different—each produces a distinct set of products or transcripts—complementarity allows the two components to work in tandem to direct and energize cellular functioning.

Redundancy refers to the characteristic shared by two body parts. The nucleus and the mitochondria of cells share the characteristic that each contains DNA. Moreover, no other organelle within the cell, except for these two, contains DNA. The repetition of DNA in these two components insures that the cell will be able to carry out its essential functions.

]OOO[

COMPLEMENTARITY AND REDUNDANCY WORK TOGETHER in our lives as they do in our bodies. When you miss out on an opportunity—you fail to get a promotion you have been hoping for, or the man you hoped to marry leaves you for someone else—you may experience depression or despair. Cellular wisdom suggests, however, that opportunities are never lost. Though one avenue to advancement may be closed, a complementary aspect of your life may benefit from what has occurred, or an opportunity may emerge in another guise. Having lost the promotion, you have more time to spend with your family; now that your relationship has ended, your former love's roommate, who has always wanted to ask you out, does so, and you marry within the year! The lesson of the body is to stay mindful and open so that you can recognize and activate complementary elements and become aware of alternate or redundant sources of abundance.

Remember Stephanie and her scholarship-winning story about her imaginary family? During her first year at college, her writing teacher, Professor Moore, invited her to join the writing workshop he sponsored. The students in the group met weekly to read aloud and discuss their work. They provided Stephanie with thoughtful feedback—the perfect complement for her creativity. She was often inspired by a character or technique in another student's story, and she used that inspiration to fuel her own creative process. The students in the group soon came to feel like family, united as they were by the same dedication to great writing and to helping each other improve.

Stephanie's new clarity about her life goals—she was sure that she wished to make writing her career—opened a whole new set of possibilities for abundance in her life. She began to actively explore various ways she might manifest her goal of becoming a professional writer. Perhaps, she thought, I'll apply for a summer internship at a magazine or publishing house, or volunteer to write a column for the college newspaper, or keep a detailed travel journal on the trip I plan to take to Spain next year—and then try to turn it into a novel. Like Stephanie, once we make the shift into abundance thinking, our lives can be filled with rich and potent opportunities.

SCULPTING A LIFE OF ABUNDANCE

Step Three: Recognizing Hidden Opportunities
Follow the method above to enter a place of quiet and relaxed reflection. When your body and mind are at ease, consider the following:

[] Allow your mind to scan your life and recall a time when you felt that you had lost out on an opportunity. Review the circumstances of this event and allow the feelings you had at the time to rise into consciousness.

[] Now gently bring your mind back from that time into your present life. Looking back from the perspective of the present, bring to mind several positive consequences that arose from the lost opportunity. Allow the feelings associated with these positive consequences to rise into consciousness.

[] Now scan your current life and bring to mind a situation in which you anticipate or fear that you may lose out on some opportunity.

[] Allow yourself to imagine freely what positive consequences may occur—what other aspects of your life may benefit, or what other opportunities may open up—if your fearful imagining does manifest itself.

[] Remind yourself that no one can see the future, and that when you stay mindful and open, any life occurrence can transform itself into a rich and potent opportunity for happiness.

A River of Abundance On Call

Once we're aware of the many possibilities for abundance in our lives, we can relax into the appreciation of the abundance that's available to us moment by moment. The body illustrates this "abundance on call," as well. Most cells in the body store nutrients not immediately needed in the form of glycogen, a compact, starch-like substance that can be changed into the simple sugar, glucose, as the body needs it. However, neurons do not store glycogen; they rely on nutrients transported to them by blood cells. If the blood supply is cut off, neurons are deprived of nutrients and may die.

Yet the advantage of accessing abundance on call as the need arises is greater than the risk. Given that neurons do not devote energy to hoarding nutrients, they are free to employ their energy for other uses. Neurons, we might say, are totally focused in the present, trusting that the abundance they need will always be available.

]0000[

THE STORY OF LISA FITTIPALDI, an artist who began painting after she became blind, illustrates how trust in the availability of abundance can help reshape a life after catastrophe. In March 1993, Lisa, forty-three years old and a certified public accountant, was driving home on the interstate from an Austin, Texas, hospital where she was employed as a financial analyst. Suddenly everything went dark and she almost collided with a truck. She lost her vision, but it returned. Three weeks later, her sight disappeared once more.

After consulting an ophthalmologist, Lisa learned that she was permanently blind because of a condition that prevented blood from nourishing her optic nerves. She could not see, even though she had corneal implants and her beautiful blue eyes shone. At a loss in the world of the blind, she spent her days in bed, depressed and crying. One day her husband threw a child's coloring set at her and said, "Do something!" She took the watercolors and painted three transparent jars.

That was just the beginning. Today, Lisa paints realistic scenes containing complex images, a delicate blending of hues and colors, lights and shadows (a sampling of her work is posted to www.blindartist.com). When she paints with watercolors, Lisa discerns the colors by the texture of the powders. She describes cobalt blue as sticky—stickier than yellow, which is light and smooth. Red is even smoother.

Lisa was told that oils would be more difficult for her to use. Rather than feel defeated, she was challenged. Now she paints with oils, even though she cannot distinguish the colors by texture or smell. Instead, her husband puts the paints on a special palette, in alphabetical order. Lisa told me that she has a photographic memory for text and graphics—she just knows where things are on the canvas.

Lisa's evolution in painting reflects how much she has grown since the day she learned she was blind. She did not choose the acceptable way to function as a blind person; rather, she used her creativity to find ways to live in a sighted world. She describes her blindness as liberating, not as cataclysmic, and she describes herself as "the most well-centered person you've ever met." When I recently spoke with her, I asked whether she was this way before she became blind. "No," she responded.

Her inspirational success as a painter—she told her story to a national audience on the *Oprah Winfrey Show,* and was featured in more than forty magazine and newspaper articles both in the United States and overseas— has also enhanced the lives of others with visual and hearing impairments. Using the proceeds from the sale of her paintings, Lisa founded and now directs the Mind's Eye Foundation, which supports the development of software for blind and visually and hearing-impaired children. She continues to serve on the Texas State Independent Living Council, appointed by then-Governor George W. Bush.

What happened to Lisa could be viewed as a tragedy, but it is also an example of genius unveiled. When she began to paint, she had no expectations. She simply turned away from self-pity and opened herself to creativity, trusting the river of abundance that's always on call to unveil the impossible. She hopes that the sighted world will learn what she learned: You can do anything if you create the opportunity.

SCULPTING A LIFE OF ABUNDANCE

Step Four: Claiming Your Abundance
In a quiet place, where you will not be disturbed, make yourself comfortable and do the following:

[] Run your mind over your current life. What resources or sources of abundance are always on call for you? For instance, is there a close friend or family member you can rely on utterly? Or, a hobby or activity that never fails to nurture you? As you consider this question, note any reactions in your body. Is there a sense of free-flowing energy? In what ways do you sometimes block your access to this river of abundance on call? As you consider this question, note any reactions in your body. Is there a tightness in your solar plexus? Does your throat feel constrained, or is there some other physical response you can identify?

[] What might you do to overcome this sense of blockage and gain free access to abundance on call? Lisa Fittipaldi had to give up her self-pity and trust her creativity. What might you need to give up? What might you need to trust?

[] Has anyone ever thrown you a lifeline or issued you a challenge, as Lisa Fittipaldi's husband did for her? What outside influences might help you gain greater access to your river of abundance?

After answering these questions, take some time to draw up a plan for gaining greater access to your river of abundance on call.

CHAPTER NOTES

1. Pascal Meier, Andrew Finch, and Gerard Evan, "Apoptosis in Development," *Nature* 407, no. 6805 (2000): 796–801.

2. Eric S. Lander et al., "Initial Sequencing and Analysis of the Human Genome," *Nature* 409, no. 6822 (2001): 860–921.

3. Eörs Szathmary, Ferenc Jordan, and Csaba Pal, "Molecular Biology and Evolution. Can Genes Explain Biological Complexity?" *Science* 292, no. 5520 (2001): 1315–1316.

4. Paula J. Grabowski and Douglas L. Black, "Alternative RNA Splicing in the Nervous System," *Progress in Neurobiology* 65, no. 3 (2001): 289–308.

5. Michael Michalko, *Cracking Creativity* (Berkeley: Ten Speed Press, 2001).

6. Richard P. Feynman, *"Surely You're Joking, Mr. Feynman"* (New York: Bantam Books, 1985); Richard P. Feynman, *"What Do You Care What Other People Think?"* (New York: Bantam Books, 1988).

7. Richard P. Feynman, *"Surely You're Joking, Mr. Feynman"* (New York: Bantam Books, 1985).

CHAPTER 7

reach out to others

RELATIONSHIPS ARE THE CRUCIBLES OF LIFE. When we are in love, we tremble with anticipation at making contact with our beloved. Operas, plays, and popular songs thrill to the celebration of love reciprocated and weep over the despair of rejection. Human beings, it seems, thrive through connection with others. Connecting energizes and fulfills us. It evokes the unfolding of who we are. Through relationship, we reach our highest potential and achieve our grandest purposes.

The same is true for the cells of the body. Though we are born with most of the neurons that serve us throughout our lives, as newborns we display few of the complex behaviors that characterize us as humans. The trillions of neurons that make up the brain and its networks are powerhouses of potential, but isolated neurons can accomplish little. Only when they connect with each other do these cells fulfill their immense potential.

Let's review the basic structure of a neuron. The cell body is the central processing factory of the cell. Molecules synthesized in the cell body support all of the cell's functions. Radiating out from the cell body are protrusions known as processes. These include a single long tail called the axon and a structure resembling the crown of a branching tree, called the dendrites. The axon is a conduit, passing along information from one neuron to another. The dendrites are the neuron's principal receiving station. They receive input along their many branches.

The axon of a neuron in the motor cortex must pass through many regions of the brain and down the spinal cord, a distance of several feet, to connect with a motor neuron in the lower part of the spinal cord. Axons travel through regions of the brain by coming together to form tracts, which function as the connecting cables between neurons. The axon of the neuron in the motor cortex (neuron 1) travels in a tract to gain access to the dendrites of the second neuron in the spinal cord (neuron 2). Neuron 1 cannot transfer information to neuron 2 until its axon extends far enough to reach the dendritic tree of neuron 2.

The information is carried by an electrical signal that travels along the membrane of neuron 1 from its cell body in the motor cortex of the brain, along the axon to the dendrites of neuron 2. The point at which the two neurons meet is called the synapse. Here, the electrical signal transforms into a chemical signal. At the synapse, the electrical signal causes the release of neurotransmitters from the tip of the axon of neuron 1. Neurotransmitter molecules diffuse across the synaptic space to bind with receptor molecules on the dendrites of neuron 2.

About a month after gestation, the first neurons are born in a developing fetus.[1] Soon after, the neurons begin to migrate, extending a process as they move toward their final sites in the brain. As the connecting cables of axons are laid down, synapses form between axons and dendrites, or axons and cell bodies. The first few years of life are a time of explosive development as axons elongate along tracts, and synapses form to link neurons throughout the nervous system. Each neuron makes an average of a one thousand connections, though some make as many as one hundred thousand connections.

Multiple neurons connect via synapses to form a functional chain, like team members in a relay race. As connections join together neurons that were formerly isolated, babies gain the ability to focus visually, talk, walk, and perform other functions. The synapses that are formed during a child's first years of life mature as the child grows. By age three, a child has more synapses then at any other point in life.[2] In fact, the synapses of a three-year-old outnumber those of an adult by 50 percent! At puberty, those connections that are less useful or robust are pruned or discarded.

In our lives, we fulfill our purposes via relationships. In the process of relating to others, we come to know who we are. Growth takes place whether our relationships are fulfilling or frustrating, long-term or temporary. It takes place whether we are relating to a spouse, parent, child, friend, neighbor, lover, or colleague. As we relate to others, we build mature and lasting connections that support us throughout our lifetimes, and prune away those transient or immature connections that do not support us.

Receiving Stations

The principal receiving station of a neuron is its dendritic tree. The size of a neuron is expanded from three to thirty times by the extent of its dendritic tree.[3] As the dendritic structure develops, tree-like processes reach further from the cell body, growing additional branches to provide future sites of connection with other neurons.[4] When fully mature, dendrites take in information along the full extent of their main branches and offshoots.

Let's look at an example. We are able to see because of a complex network of neurons. The first elements of the visual system are the specialized photoreceptors, or detectors of light—the rods and cones of the eye's retina. The photoreceptors in the retina communicate the information they receive to neurons within the retina. Information flows from these retinal neurons, via the optic nerve, to a collection of neurons in the thalamus region of the brain. Thalamic neurons relay the information they receive from the optic nerve to the visual cortex at the back of the head. Thus vision requires that information be transferred through three processing sites: the retina, the thalamus, and the visual cortex. Only when the information arrives and is processed in the visual cortex do we actually see.

Newborns have very poor vision, estimated at 40 percent of the acuity of adults.[5] Moreover, newborns can't discriminate color well. Researchers have found, for example, that newborns cannot differentiate blue, green, or purple from white, nor yellow from red. The poor sight of newborns reflects the immaturity of the retina's neurons and those of the processing sites along the visual pathway.

Between birth and six months, a fivefold increase in visual acuity occurs. By three months, infants begin to make color discriminations. This improvement echoes the maturing cones' ability to detect progressively finer detail, and the maturing of neurons in the processing sites. By age six, a child's vision equals that of an adult. This final stage of visual improvement mirrors the growth and expansion of the dendritic trees of neurons in the visual cortex, the increase in the number of synapses on these neurons, and the maturation of the visual pathway.[6]

Neurons have dendritic trees with a variety of shapes, depending on their function. For instance, neurons in the thalamus have dendritic trees that are round and symmetrical, with branches extending equally in all directions. This shape makes sense, because these neurons receive and redirect input coming from a host of different pathways to appropriate processing sites in the cortex.

Many neurons in the spinal cord, on the other hand, have dendritic trees that are shaped like a broom, one set of bristles extending up and a second extending down. This shape also makes sense, because these neurons transmit information up and down the spinal cord.

]0000[

INTEGRATING INFORMATION THAT WE RECEIVE from a rich variety of connections is as important to our lives as it is to our cells. Like the neurons of our visual system, we extend our receptive fields into the world via our families, work, interests, and activities. The sum of all our connections permits us to see the world as neurons do—from many different perspectives. If all of our connections were focused in only one direction, our world would shrink to one dimension.

Remember Cassandra, the disillusioned criminal lawyer we met in chapter 1, who wanted to make a living raising horses? With the help of a financial planner, Cassandra developed a strategy to switch careers. As her business matured, she reached out to other successful horse breeders. It was easy for Cassandra to relate to these people because they shared her love of horses. Each person she met in the business introduced her to others—specialists in veterinary medicine, trainers, stud farm owners, auctioneers, architects, and builders. Within a few months, Cassandra had generated a wide network of people who were willing to help her with various stages of her developing business. Generating the equivalent of an expanded and functional dendritic tree, widely dispersed but focused on her particular interest, Cassandra received useful information from people with varied expertise, which helped her put her plan into practice.

Being receptive to input from others is especially important when we're in the midst of change. When I was in the process of deciding whether to leave my life in academia, I felt stripped of nearly all of my familiar connections. Colleagues just shook their heads when I tried to talk to them and told me, "You just need a break, Joan. After your sabbatical, you'll feel better about things." Rather than remaining in that solitary place, before I began my sabbatical, I signed up for a writing workshop on the beautiful island of Monhegan, off the coast of Maine. There were ten people in the workshop, all of us struggling in one way or another with the questions I was asking myself: Who am I? What's next for me? One woman, in particular, caught my interest. Exuberant, earthy, and humorous, Jeannie always told it as it was. After the workshop ended, Jeannie and I remained friends.

As the months of my sabbatical year passed, I became more agitated about deciding what to do. It would be so easy to slide back into my familiar set of connections at the university. During a visit to Boston from her home in New York, Jeannie listened intently as I described what I might do when I returned to the university. "Joan," she said after I had exhausted myself, "when you speak about your old life, there is no excitement in your voice and no glimmer of joy in your eyes. Maybe you don't want to hear this, but it's clear to me that no matter what you say, you don't really want to go back."

Jeannie's words hit me like a punch in the stomach. I knew she was right, although I had never allowed myself to articulate the truth about how I was feeling. Her input gave me the opportunity to face my hopes and fears squarely. I knew then that I could not make the easy choice and go back to the life I had left. In order to grow, I had to continue to seek input from fresh voices. I hired a personal coach, enrolled in more writing workshops, and began to attend personal growth seminars. The more people I met and the more new connections I formed, the more support I felt for my resolve to move forward on my new path.

It's interesting to consider the shape of your own pattern of relationships. Do you form relationships primarily with people who are older or somehow above you, or with those who are younger or in some way below you, as do neurons in the spinal cord? Do you form simultaneous connections with many kinds of people and have many types of relationships, as do the thalamic gateway neurons? The following exercise can help you think about your receiving station and what it shows about your ability to receive input from others.

YOUR TREE OF RELATIONSHIPS

Find a comfortable space. Relax, breathe deeply, and rest in the energy that flows into you moment by moment. When you've reached a state of quiet and calm centeredness, respond to these suggestions and questions:

[] Draw and label the branches of the dendritic tree of your current and most important relationships, keeping in mind the various shapes of dendritic trees you have read about. As you draw, consider the qualities of your various connections. Long branches, for instance, might indicate connections with people who are geographically separated from you, or who are much older or younger than you.

[] What does the shape of your tree of relationships tell you about yourself?

[] Was there a time in your life when your tree of relationships had a different shape? If you wish, draw the tree that existed at some earlier point in your life.

[] Was there a time in your life when you were particularly happy or particularly sad? If you wish, draw the tree that existed at one of those times.

[] How would you like to change the shape of your current tree of relationships? If you wish, draw your ideal relationship tree.

[] Take a moment to write about what changes you would like to make, and list ideas for how you might make these changes.

Making Appropriate Contact

The second element vital to connections between neurons is the axon, the long tail that sends the signal. As the nervous system develops, axons explore their environment extending over short or long distances to find the right targets for their transmissions, the dendrites of receiving neurons.

Like heat-seeking missiles that bypass potential targets before zeroing in on the right ones, axons find their targets with precise specificity. They deftly avoid contact with many a neuron, or muscle, gland, or blood vessel as they zip toward the appropriate path sites. Molecules at the tip of the growing axons guide axons through this process. In fact, exploring axons are guided along appropriate routes and toward desirable targets by a series of molecules that act as tags.

For instance, a group of neurons essential to normal reproductive functioning is born in the nose and migrates into the brain as the nervous system develops. Although the specific tags have yet to be identified completely, it appears that not one, but many tags guide the neurons out of the nose and into the brain.

Among the molecular tags that help axons find the right places to link up are a group of proteins called cadherins. These proteins occur on the growing tip of an axon and on potential connection sites on the dendrites of other neurons. As the axon approaches the dendrite, if a match is made between the cadherins on the axon and the cadherins on the dendrite, a synaptic connection forms. More than eighty different cadherins have been identified by researchers.[7]

]000[

THE ROLE OF CADHERINS and other guiding molecules in bringing together axons and dendrites is a useful model to consider when thinking about our own life choices. Many people, even those in middle age, may still be puzzling about what they want to be when they grow up. What they're looking for is that click of connection, that synapse that forms when their talents, interests, and personal qualities find an appropriate target.

Sometimes, like exploring axons, we need to look far away from our familiar surroundings to find the right place to link up. I think, in this regard, of one of my heroes, Albert Schweitzer. A deeply religious man, Schweitzer determined at a young age that by the time he was thirty, he would find some way to devote his life to the service of humanity. One morning, while leafing through a magazine, he noticed an article with the headline, "The Needs of the Congo Mission." It was written by Alfred Boegner, the president of the Paris Missionary Society. Boegner wrote that the mission did not have enough people to carry on its work in the Gaboon, the northern province of the Congo colony. As Schweitzer recounts in his autobiography: "I finished my article and quietly began my work. My search was over."[8] Though he had already earned a Ph.D. in philosophy and had studied theology and music, Schweitzer was determined to go to medical school so that he could serve the people of Africa. When he finished his medical studies, he raised enough money to establish and operate a small hospital in Africa to serve the Paris Mission Society. He and his wife, Helene Bresslau, who had studied nursing to assist him, then traveled to Africa at their own expense. The public health conditions they found were awful. Schweitzer's first clinic was an old chicken coop near his living quarters, from which he began treating the terrible diseases of the Congo—leprosy, malaria, dysentery, heart disease, and pneumonia.

Yet, as Schweitzer tells us, his life in Africa was characterized by immense joy and satisfaction. The guiding tags of Schweitzer's desire to serve and the overwhelming needs of the people of Africa linked the two into a functional and fulfilling relationship. "The important thing is that we are part of life," Schweitzer wrote. "We are born of other lives; we possess the capacities to bring still other lives into existence. In the same way, if we look into a microscope, we see cell producing cell. So nature compels us to recognize the fact of mutual dependence, each life necessarily helping the other lives which are linked to it."[9] By placing himself in the service of others, Schweitzer forged the relationship that brought meaning and purpose to his life.

Sometimes the relationships that bring us joy during one part of life fade as time goes on. For many years I was happy teaching medical students about the nervous system and conducting my own research. I received tremendous satisfaction from seeing students grasp the concepts I was teaching and use the computer-assisted programs I designed to help them understand how the nervous system operates. Later, as an administrator, I channeled that satisfaction into the creation of a Multimedia Resource Center to help students with many aspects of their medical studies.

Over time, however, and with financial constraints restricting what I could create, the click of connection faded. There was so much more I wanted to do. Inspired by Schweitzer, I wanted to share with students what I believed about the essential link between medicine and service. But the appropriate vehicles for expressing this desire were not available. Increasingly, I felt that my profession was activating only a part of my wholeness. I longed for a new set of connections that would allow me to express everything I was and everything I knew.

My explorations since leaving academia have led me to a whole new set of relationships. The most important one is with myself. I now spend a considerable amount of time alone, reflecting and writing. My work with individual coaching clients is similarly fulfilling, as the connections I create help people transform their beliefs and attitudes. And, yes, I'm back in medical schools as well, but now I give workshops that facilitate the search for the deeper meaning of the profession. Our tags, it seems, can guide us into relationships and out of them, toward new and more satisfying connections.

What role, you might be wondering, have tags played in the formation of your relationships? Do you have tags that attract or repel others? Do those with whom you wish to connect have tags that facilitate or impede your ability to connect them? And, if you do have tags, what do they consist of— ways of behaving? Talents or interests? Personality traits? Inner qualities? How have these tags guided the formation of your most important relationships? The exercise that follows can help you explore these questions.

THE TAGS GUIDING YOUR CONNECTIONS

Find a comfortable space, relax, breathe deeply, and rest in the energy of your body. When you've reached a calm interior state, consider these questions and suggestions:

[] What are your tags—specific ways of behaving, talents, hobbies and interests, personality traits, inner qualities, beliefs?

[] Choose one or several of these tags, and think about the connections you have made guided by these tags.

[] Which of these connections bring you joy and fulfillment? Why? Which tags that facilitated the connections continue to be important to your well-being?

[] If one or more of your connections is no longer fulfilling, why is this so? Have the tags that facilitated these connections lost their importance? If so, why?

[] Taking a broader perspective, think back to a time in your life when you were particularly happy. Which tags were guiding your relationships at this time?

[] Thinking back to the Tree of Relationships you drew earlier in this chapter, which tags were active in helping you form your most important relationships?

[] Are there any tags on your list that you have not explored fully?

[] What might you do to explore the new connections these tags might facilitate?

As we have seen, neurons make connections both by reaching out with their axons and by receiving input along their dendrites. As you continue to think about your connections and what has guided them, consider to what extent you are skilled at both reaching out and receiving in your relationships. If you find that your relationships are characterized more by one mode of relating than the other, ask yourself what you might do to balance these two essential ways of connecting. What steps might you take to strengthen your ability to both reach out and receive?

Immature Connections

The first connections formed between axons and dendrites are immature—that is, they are neither electrically nor chemically precise. As the nervous system develops during gestation, these immature connections become functional and accurate.

Here's an example. Before a baby's eyes open and before the eyes have begun to respond to light, neurons in the retina spontaneously begin to fire.[10] Then, neighboring groups of retinal neurons fire together, synchronously sending impulses to the thalamus, on either side of the brain. As groups of neurons across the retina fire together, a wavelike pattern emerges, in which periods of quiet alternate with waves of random but synchronous activity. Because the eyes are not open and no visual images are being sent to the brain, what is the purpose of this neural activity?

The answer is that the firing of the immature circuits connecting the eye and the brain is actually modeling the connections between the retina and the thalamus. For us to see, axons reaching the thalamus from the retina of each eye must connect in a precise, layered pattern to produce a map of the retina. Moreover, the axons from each eye must connect to different layers.

When the axons of the retinal neurons first reach the thalamus, their branching endings reach into all layers. During the wavelike pattern of firings, axon branches in inappropriate layers retract, while those in appropriate layers expand within the layer. This modeling leads to the adult configuration of the thalamus, in which each eye sends input to different layers. The sequential, coordinated activation of retinal neurons is essential to developing the mature connections that make sight possible.

]OOO[

OUR INITIAL ATTEMPTS AT CONNECTING with others may be likewise immature, and they may yield unsatisfactory and unpredictable results. Justin, age fifteen, had a crush on Juliet, a girl in his sister Mary's senior high school class. Every time he was near her, Justin's gangly frame seemed to lengthen, and his arms and legs twisted into awkward configurations. His brain seemed to freeze, and the words he stuttered always sounded too loud and sharp to his ears. Amazed at the girl's ability to unnerve him, Justin longed to reach out and connect with Juliet, but he had no idea how to do it.

When Justin's brother, Mark, came home from college for a visit, Justin got up the courage to ask him about it. As Justin began to tell Mark about Juliet, Mark threw back his head and laughed. For a minute, Justin was afraid Mark was making fun of him, but then Mark's blue eyes began to radiate warmth and understanding. Mark told Justin about some of his own clumsy attempts to connect with girls. As they talked, Justin felt better and better. He understood that it would take time and many attempts before he felt confident about approaching a girl he liked.

We adults are not immune to the same turmoil about making connections. When we meet someone who throws our feelings into disorder, we may become like awkward teenagers again. This upset may be the call of expansiveness, an indication that it is time to widen the circle of our relationships. We might do well, in such circumstances, to follow the lead of our cells. Set a slow tempo in synchrony with the other person, alternate periods of activity with periods of quiet and withdrawal, and allow impatience to be present without letting it drive the interaction. Giving a relationship time to unfold naturally, without trying to manipulate it, is truly a discipline of maturity.

When he went to work that Thursday, Tobias, a good father and a caring husband, did not expect to meet Angela. Particularly, he did not expect to feel like a dumbstruck kid again—certainly not at age fifty-four! Angela was a new employee at the graphic design company where Tobias worked. When she approached him for advice on a project, Tobias seemed to forget everything he'd learned during thirty years in the business. He was stunned by his reaction. He had no intention of pursuing an affair with Angela, but he couldn't get her out of his mind, especially since she continued to approach him for advice about her projects.

Tobias knew he needed help. He decided to talk to his brother, Jonathan, during one of their weekly lunches. He began the conversation by asking his brother, "Have you ever been attracted to another woman?"

Ah, so that's what this is what this is about, Jonathan said to himself. He knew it was the time for absolute honesty. "Sure," he replied, "but I love my wife and made sure not to get involved."

"What did you do? Quit your job?"

"Nothing as serious as that," Jonathan replied lightly. "I took the advice of my neighbor. You remember that nice old guy who lives next door? One day, over the back fence, I told him about this woman, Elizabeth, at work I had been thinking about a lot and asked him how he would handle the situation. He gave me the best advice I'd ever heard. 'Introduce her to your wife,' he told me."

"That was the last thing I would have thought of doing!" Tobias said with a grin. "So, did you do it?"

"Yes, and it did the trick."

Jonathan told Tobias that putting his relationship with Elizabeth into the context of his being a married man changed the dynamic of their relationship. The attraction disappeared, and over time, Elizabeth became a friend of both Jonathan and his wife. Later, when Jonathan's wife became ill, it was Elizabeth who helped him out the most.

"Relationships can't be put in boxes," Jonathan said. "As human beings, we are meant to be attracted to each other and to reach out for connection. Sometimes the resulting relationship doesn't fit neatly into a category. All I can tell you, Tobias, is that my wife and I have been very grateful for Elizabeth's friendship all these years. Running away from the attraction would have cheated us all of very meaningful friendships."

MAKING NEW CONNECTIONS

Use the techniques you know to reach a calm and centered state. Then consider these questions to assess your level of openness to exploring new relationships:

[] When you meet new people, how do you respond? Do you consider yourself to be shy? Friendly? Self-conscious? Relaxed and open? Curious?

[] Do you ever allow yourself to explore relationships that are different from those you have with your usual friends?

[] Have you ever decided not to explore a relationship because the person did not fit into one of your usual categories of associates?

[] Have you ever invited someone who doesn't fit into a neat category to join your family and friends for a celebration?

[] Is there someone in your life now with whom you might consider exploring a new connection? What first step might you take?

]OOO[

CULTIVATING THE ABILITY TO EXPLORE new relationships while maintaining your authenticity can enrich your life. Not all relationships fit into the categories that society has defined. For example, Tobias taken back by his attraction to Angela, initially thought his only alternative was to turn away from her, valuing his marriage. His brother Jonathan suggested that Tobias introduce Angela to his wife, an alternative he did not consider. All of their lives could be richer for this expansive and authentic alternative. Exploring does not mean committing, nor does it mean being unfaithful to existing friends and family members. It means staying centered and true to yourself while being willing to open to new connections.

Shaping Mature and Functional Relationships

As we have seen, the neural circuits that make sight possible begin to take shape before the retina is able to respond to light. But significant, further shaping occurs once the photoreceptors become active. Each time light reaches a photoreceptor and causes it to fire, a signal is sent through the neurons in the retina, along the optic nerve to the thalamus. Other neural pathways send the signal to the visual cortex at the back of the head, where sight actually takes place.

As in all neural transmissions, the signal passes from axon to dendrite at junctions called synapses, but the axons and dendrites do not actually touch. Chemical messengers called neurotransmitters carry the signal from the axon of one neuron to the dendrite of the next, across a fifty-nanometer space. This process of repeated sending and receiving, shapes and matures the circuits by stabilizing the synapses between neurons along the path from the eye to the brain. That explains why babies learn to see better by looking at things. Without sufficient visual stimulation to shape the circuits, a baby's sight may fail to develop fully.

]000[

WHAT CAN WE LEARN FROM THIS MODEL about our human relationships? For one thing, it's clear that frequent and appropriate interactions are needed to stabilize and mature our connections to others. When one member of a married couple, for instance, works long hours and misses dinner most nights, or otherwise withdraws from frequent and meaningful interaction, the circuits linking the pair may begin to break down. Remember the car salesman, Frank, who was so intent on succeeding as a manager that his link to his wife, Elaine, and their child withered? We might say that because of lack of stimulation, the circuit linking Frank and Elaine deteriorated so significantly that it could not be rejuvenated, as happens when an axon branch retracts.

Equally important is that the stimulation of the circuits that link us to others be appropriate. Justin's awkwardness over approaching his sister Mary's friend, Juliet, began to fade when Justin discovered that he and Juliet shared an interest in classical music. When Justin's father gave him three tickets to the local symphony, Justin asked Mary whether she and Juliet would like to go with him. The evening was a great success. Justin had studied the cello. Juliet adored the violin. Conversation flowed as they discussed their favorite performers and composers. It was almost as if the music set up a chemical reaction, carrying communication and good feeling effortlessly across the space between them. They even made plans to attend an upcoming concert featuring a violinist and a cellist they both admired.

In time, and because of many such conversations and interactions, Justin and Juliet became good friends. Though their relationship never budded into romance, Juliet introduced Justin to many of her friends, including Elizabeth, with whom Justin did become romantically involved. The chemical transmitters and receptors between Elizabeth and Justin generated a different kind of communication than the one between Juliet and Justin. But because Justin had gained experience in communicating and connecting with Juliet, he was able to develop his relationship with Elizabeth. Time and practice matured his ability to relate to others and led to stable connections with both women.

FORMING MATURE AND STABLE CONNECTIONS

Find a comfortable space and breathe deeply. Imagine the billions of neurons connecting to each other via trillions of synapses throughout your body, effortlessly releasing neurotransmitters and receiving responses from specific receptors. This thriving communication is the substrate of life—vital, pulsing signals linking our various parts. Recognize that neurons, in their continual, ongoing linkages, can teach us about forging relationships. Rest in this flowing energy of communication and allow any tension you may feel to relax. When you've reached a calm state, reflect on the following situations and questions:

[] Recall the stages of relating within your body: reaching out as dendrites do in development, growing in response to attraction as axons do, forming immature connections, and maturing these connections through the release of neurotransmitters and their binding with responsive receptors.

[] Bring to mind relationships that typify each of these stages in your own life—occasions of reaching out, responding, grappling with immature connections, and—finally—experiencing interactions that have forged mature and stable bonds between you and others.

[] What relationships in your current life are most critical to supporting your essential self and maintaining your ability to relate meaningfully to others?

[] What specific steps are you taking to stabilize, maintain, and strengthen these connections?

[] Are there any aspects of the way you relate within these connections that you wish to change? For example, are your interactions consistent? Are they appropriate? Do you engage in behaviors that disrupt the flow, such as repeating gossip or continuing long-held grudges? If it feels appropriate, resolve to make needed changes that enhance the flow of communication and interaction between you and your significant others.

[] Remember that, like neurons, we cannot fulfill our life purposes alone. What undeveloped relationships do you wish to strengthen to help you fulfill your life purpose? For instance, is there someone at work, in your neighborhood, or in an organization to which you belong who you would like to know better?

[] What might you do to develop this relationship? What common interest might serve as the neurotransmitter to help you stimulate and mature this connection? Articulate a step you might take and commit to that action.

[] Now turn your thoughts to your relationship with yourself. Allow yourself to consider whether you are making the time and taking the appropriate actions to fulfill your own needs in a stable and mature way. What are you doing on a daily basis to relate to yourself: Making time for meditation or reflection? Engaging in hobbies or creative projects? Exercising or pursuing other health-related activities? Participating in recreational activities? What more might you do? Articulate a step you might take and commit to that action.

[] Finally, ask yourself what you are doing to participate in relationships that help unfold the lives of others. Who might be reaching out to you for connection? What might you do to respond to their initiative?

]□□□[

THERE IS NO NEED TO FEAR initiating, sustaining, and maintaining relationships. Your interior teacher is expert in this domain. When you are troubled by your relationships, turn to the wisdom of your cells for guidance. Relationships are the way we unfold ourselves and bring the wonder of what—and who—we are to the world.

CHAPTER NOTES

1. Barbara Clancy, Richard B. Darlington, and Barbara L. Finlay, "Translating Developmental Time Across Mammalian Species," *Neuroscience* 105, no. 1 (2001): 7–17.

2. Janet A. DiPietro, "Baby and the Brain: Advances in Child Development," *Annual Review of Public Health* 21, no. 1 (2000): 455–271.

3. Hollis T. Cline, "Dendritic Arbor Development and Synaptogenesis," *Current Opinion in Neurobiology* 11, no. 1 (2001): 118–126.

4. Angel Acebes and Alberto Ferrus, "Cellular and Molecular Features of Axon Collaterals and Dendrites," *Trends in Neurosciences* 23, no. 11 (2000): 557–565.

5. Daphne Maurer and Terri L. Lewis, "Visual Acuity: the Role of Visual Input in Inducing Postnatal Change," *Clinical Neuroscience Research* 1, no. 4 (2001): 239–247.

6. Helen J. Neville and Daphne Bavelier, "Effects of Auditory and Visual Deprivation on Human Brain Development," *Clinical Neuroscience Research* 1, no. 4 (2001): 248–257.

7. Barbara Ranscht, "Cadherins: Molecular Codes for Axon Guidance and Synapse Formation," *International Journal of Developmental Neuroscience* 18, no. 7 (2000): 643–651.

8. Albert Schweitzer, *Out of My Life and Thought* (Baltimore: The Johns Hopkins University Press, 1998).

9. Albert Schweitzer, *Animals, Nature and Albert Schweitzer*, ed. Ann Cottrell Free (Boston: The Albert Schweitzer Fellowship, 1982).

10. Anna A. Penn and Carla J. Shatz, "Brain Waves and Brain Wiring: The Role of Endogenous and Sensory-Driven Neural Activity in Development," *Pediatric Research* 45, no. I (1999): 447–458.

PART II

living

expansively in

community

WHAT IS COMMUNITY?

The perspective in the second half of this book shifts from a focus on individual cells to the ways cells work together in systems. As we look at how the cells in our body operate in systems, we are seeking guidance for how we might get along better with each other in our human communities—in families, neighborhoods, cities, businesses and organizations, nations—indeed, the entire planetary human family.

During the billion years following their emergence, cells aggregated, diversified, coalesced, and integrated into multicellular complexes. Three strategies were central to the success of this evolutionary process: specialization, cooperation, and integration. As we will see, these fundamental principles are as important to creating and sustaining vibrant human communities as they are to the functioning of body systems.

Specialization focused the life force pulsing in the interior of cells in specific directions to give rise to many diverse types of cells. Groups of specialized cells then aggregated into tissues, organs, and systems. For instance, germ cells evolved to devote themselves exclusively to reproduction, which freed other specialized cells to concentrate on fulfilling other body functions. Previous to this development, single-celled organisms were stationary when they were reproducing, and mobile when they were not. Thus the lives of such creatures alternated between mobile and stationary phases. Specialization began the process of freeing organisms from this biological imperative.

The differentiation of specialized cell types also allowed aggregates of cells to grow larger and join together to form multicellular organisms composed of many different groups of specialized cells. This change allowed for the evolution of the nervous system. As organisms increased in size, neurons diversified, specialized, migrated, and organized into unified structures called nuclei and ganglia. A cluster of neuronal cell bodies is called a nucleus when it resides within the central nervous system, the brain, and the spinal cord. Similar clusters of neuronal cell bodies outside of the central nervous system are called ganglia.

The neurons of a nucleus or ganglion cluster together to serve a specific function, such as sensing conditions in the external environment and transmitting that information to the brain. During the process of evolution, sensory cells once on the periphery of organisms migrated inwardly and took up positions alongside the spinal cord. Such an evolutionary change enables extensions of the sensory neurons in these ganglia, called dorsal root ganglia, to carry sensory information into the central nervous system, where the information is transmitted to the thalamus. A group of neurons within the thalamus receives this information and relays it to the cortex, where the sensory experience is made conscious.

The visual system is an example of how specialization of cells within the nervous system enhanced the capacities of an organism. The photoreceptors within the retina allow specialized cells to be exquisitely sensitive to light and color. Specialized neurons in the thalamus do not detect light and color directly, but they are excellent integrators of information received from the retinal neurons. Neurons in the occipital cortex assemble information from both eyes and weave it together to generate a visual map of the world. In that way, specialization generates diversity. If all cells specialized in the same way, no new capacities could emerge. In the expression of diversity, the genius of each species emerges.

Cooperation necessarily flows from specialization. In order for specialized systems to work together to achieve common aims, complex highways and networks evolved to allow cooperation between specialized components and mobilize their influence for higher functioning. Consider, for example, the vascular system. This grand network of rivers and tributaries though which the blood flows, carries life-giving molecules to the cells of the body, including nutrients absorbed from the foods we eat by the digestive system and oxygen gathered from the air we breathe by the respiratory system. Without such a vast distribution system, multicellular organisms and their complex organs could not be sustained.

We discussed previously how red blood cells are continuously generated within the marrow of bones in the skeletal system to supply the vehicles for transporting oxygen and carrying away waste. The force of the beating heart, the engine of the circulatory system, propels the movement of these blood cells. Only synergistic cooperation between the vascular system, the skeletal system, the digestive system, the respiratory system, and the circulatory system—each a specialized element—gives us the ability to sustain a functioning multicellular body.

No system of the body stands alone. Each must integrate with other cellular systems, working together to achieve a common goal. As the muscular system increased in complexity, for example, intricate arrays of longitudinal and transverse cables developed. These muscle cables, when attached to the bones of the skeletal system and stimulated by expanded neural networks, made precise and complex movements possible, including walking upright. Other physiological functions, such as reproduction, require integration among many different systems. In order for a mammal to reproduce, the brain and pituitary, both of which have other functions within the body, must integrate with reproductive organs, such as the gonads and uterus, which are specifically designed for reproductive purposes. The brain secretes a hormone that stimulates the secretion of another hormone by the pituitary. This hormone, in turn, stimulates an egg follicle in the ovary to mature and expel an egg, or it stimulates the testes to generate and ejaculate sperm. The fertilized egg implants in the uterus, a placenta develops, and the growing embryo and fetus is nourished. The functional integration of these various organs and systems is necessary to produce a viable offspring.

As you can see, the human body is a complex and interconnected weave of specialized, cooperative, and integrated cellular systems. The boundaries between organs within a system, such as the reproductive system, and between systems, such as the muscular and nervous systems, are fluid and dynamic. Various systems must interact in order for the body to function. For instance, it is the synapses, or communication interfaces, between the neurons and muscles that allow us to move. When a neuron secretes a specific neurotransmitter from the tip of its axon, it activates receptors on the muscles that signal the muscle fibers to contract. Similarly, tendons—the connections between muscle and bone—stabilize the muscle and help the bone resist the pull of gravity, giving us the ability to stand, walk, and perform a wide range of physical motions. The nervous system, muscular system, and skeletal systems continuously integrate their functions by feedback and communication. The boundaries between systems function not to isolate systems, but to preserve each system's integrity so that communication can flow smoothly between them. In this way, the body, as teacher, shows us how widely diverse systems, sending and receiving various kinds of signals, operate as a harmonious whole.

]0000[

THESE SAME THREE GENERAL PRINCIPLES of specialization, including diversity and the emergence of genius; cooperation, realized through communication; and integration, in which various systems contribute resources toward achieving a common goal, are essential to creating and sustaining vibrant human communities. Like cells clustered in organs and systems, people group together in communal complexes—family groups, clans, tribes, city-states, and nations. Just as the growth of physiological systems evoked new capacities within cells, the potentials of each individual within the group are actualized when humans operate within the context of larger social groups.

Communities become and remain vibrant and flexible as long as their activities are aligned with their central purpose for being. This principle is applicable to the communities where we live and where we work, as well as to nonprofit organizations such as schools, churches, and universities. Communities that keep their central mission or purpose clear and at the center of their activities, thrive. Activities that expand the original focus of the community into new areas or applications stimulate growth, while those that distort or deviate from the core purpose of the community lead to fragmentation and disharmony. Chapter 8 explores the importance to communities of maintaining core values.

In addition to being members of communities based on physical proximity, we also belong to communities based on common interests, such as shared political views, recreational activities, spiritual values, or work interests. Modern technologies of communication and travel allow us to expand our communities beyond the limitations of physical proximity. Such communities of affinity serve to amplify the reach and scope of each member's ideas and feelings. It is no exaggeration to say that our communities are the levers by which we move the world. Chapter 9 focuses on the ways communities promote amplification, allowing individuals to boost the signal of the emotional brain.

Each community to which we belong also gives us the opportunity to interact, to send and receive feedback, both within the community and between the community and other individuals or groups. Taking time to listen to feedback helps communities maintain their balance, and interact smoothly and effectively with others. In chapter 10 we consider how various forms of feedback help communities function. We also focus on clear and accurate communication. If signals are distorted or other types of miscommunication occur, the health of a community and its members can be threatened. In chapter 11 we explore the principle of interdependence, how providing and accepting support from others is essential to achieving communal goals.

In chapter 12 we look at the responsibility of communities to protect their weakest members, and how doing so promotes the health of the whole.

In chapter 13 we explore the value of diversity to community. In the body diversity emanates from specialization. Likewise, thriving communities develop and are sustained when specialization combines with cooperation, communication, and integration. Diversity gives our communities resiliency, making us capable of adapting and responding to many different circumstances.

In chapter 14 we explore how synergy, or cooperative action, allows individuals within a community to access the genius of others. As is true with body systems, a community achieves its full potential by actively involving all of its elements. As we will see, harmonizing diverse elements within a community expands what is possible for individuals and for the group as a whole. Like the body, a community is an energy field in which cooperation among diverse elements is necessary to produce results.

The world is dangerous place these days, as we all are aware. Tension between the fragmented components of the human family is as dangerous to the health of the planet as it is within the body. In troubled times, we often look to our leaders to clarify and express the central purpose of our community, to keep it always before us. But as members of an ever-changing community, we cannot be passive and depend on our leaders to be the sole articulators of communal goals. Now, more than ever, we must be active, responsible contributors to communal life, committed to interaction and clear communication, interdependence, and compassion for the weakest members of the human family. The body can be our teacher, demonstrating principles through which we can become more responsible leaders and followers, world citizens and world lovers, powerfully shaping our communities to make them cohesive and stable.

When we follow the body's lead in this way, we each have the opportunity contribute our unique note to the universe's cosmic song.

CHAPTER 8

discover your core values

HUMAN VALUES, SUCH AS LOYALTY, SERVICE, friendship, and success, circulate through society in much the same way that chemical substances, such as hormones, circulate through the body. Values provide the energy that makes things happen at every level of society, in our neighborhoods and work places; in our churches, schools, and universities; and in dealings between nations. In this chapter, we explore what values are, how we acquire them, and how they motivate our feelings and actions. As we will discover, our families and communities—even the corporations in our communities—become and remain vibrant as long as their activities are aligned with their core values, their central purpose for being.

Let's begin by looking at a simple example of a substance that is carried by the blood vessels as it circulates through the body. After you eat a meal, the level of glucose, the sugar molecule that carries energy to the cells, rises in the blood. Specialized cells in the pancreas detect the rise in glucose levels and respond by secreting insulin, the hormone that helps the cells use glucose. Insulin enters the blood stream and circulates freely through the body. Like other signaling molecules, insulin works like a key to unlock cellular functions. Cells that recognize insulin lock onto it through specialized receptors and respond accordingly. For instance, cells in the liver lock onto the insulin and use it to convert glucose into glycogen, a form of energy that can be stored for later use.[1] Only those cells with matched receptors can lock onto the circulating insulin and make use of it.

Values travel through society in much the same way that hormones circulate through the blood stream. Messages that convey the value of success, for example, are communicated to us as we grow, through the advice we hear from parents and teachers. These messages are reinforced by political speeches, advertising, movies, television, and the Internet. In multiple ways, through words and images, we are urged to make a difference, be a winner, and "be all that you can be." Much as cells with matched receptors lock onto insulin, people who are receptive to these messages lock onto the value of success and make it part of who they are. Once the value becomes internalized, it influences the choices people make and motivates their behavior, sometimes at an unconscious level.

In this chapter, you'll have the opportunity to discover the values you have internalized, and to explore how they influence your choices and motivate your behavior.

Revealing Hidden Values

A value is an internalized belief that transcends specific applications and functions as a principle to guide behavior. In general, we act to preserve what we value and discard what we do not value. While it is relatively easy to identify the principles that govern the interactions between signaling molecules and receptive cells within the body, the influence of human values on behavior is much more complex. Often, we are not aware of the values we hold until circumstances cause us to act in a particular situation.

Let me give you an example. Recently, I had to decide whether to invite my cousin Arnold to a family reunion I was organizing. Arnold had just been released from prison, where he had spent two years as a result of a drunken driving conviction. Opinion among the members of my extended family was split about whether to include Arnold in the party. Some parents of young children were concerned about whether Arnold would be a bad influence, and whether or not they could trust him to be a loving uncle to their children. Some older relatives wondered whether they would feel comfortable about welcoming someone who had been in prison back into the family.

When I called Arnold to tell him about the reunion, he told me how difficult his life had become because he could no longer drive a car and had very little money. The conversation opened a whole new set of value-laden questions: What kind of help, if any, should the family offer to Arnold? Should someone be asked to pick Arnold up and drive him from his home in another state to the reunion? What should be said to the children in the family about Arnold's experience? In discussing these issues with members of my family, it became clear that as a result of this situation, values normally hidden were exploding to the surface.

Research psychologists actively studying the effects of hidden values on human behavior can help us understand what is going on beneath the surface of such personal value conflicts. In one study, Shalom H. Schwartz, a professor of psychology at Hebrew University in Jerusalem, administered a values questionnaire to more than twenty-five thousand people in forty-four countries between 1988 and 1993.[2] People responding to the questionnaire included elementary, high school, and university teachers; adolescents; university students; and adults in various occupations. The questionnaire consisted of a list of values presented as thirty nouns, such as *social power, freedom,* and *creativity,* and twenty-six adjectives, such as *forgiving, broad-minded,* and *ambitious.* Respondents were asked to rate the value words on a numeric scale, from those of supreme importance, to those deemed not important, to those considered "opposed to my values."

In analyzing the results of this study, Schwartz concluded that values take the form of conscious goals that allow us to cope with three sets of needs common to all individuals and societies: first, the needs of individuals as biological organisms; second, the need for coordinated interaction between individuals; and third, the need for the smooth functioning and survival of groups. Schwartz identified ten specific values that help individuals and groups meet these needs. They are power, achievement, pleasure, stimulation, self-direction, universalism, benevolence, tradition, conformity, and security.

Schwartz suggested further that a dynamic relationship exists among these ten values, which allows people to apply values in an integrated way when situations bring values into conflict. No matter which culture he studied, values were organized into sets of opposing principles. Across cultures, one group tends to esteem openness to change (for example, a high-value rating for self-direction and stimulation), while an opposing group prizes conservatism (a high-value rating for tradition, conformity, and security). Similarly, one group within society tends to uphold self-enhancement, which translates into the specific values of power and achievement, while an opposing group honors self-transcendence, which leads them to rate universalism and benevolence as very important. Because the values revealed by the study transcend nationality, age, and occupation, Schwartz concluded that these ten values and their systems groupings are basic to the human condition.

How might Schwartz's study shed light on the values conflict that erupted in my family over cousin Arnold? It seems clear that those people, like myself, who value self-transcendence, favored inviting Arnold to the reunion and offering him family aid and support. Those who wished to exclude Arnold lined up on the side of conservatism, and its values of tradition, conformity, and security. Those same conservative family members would likely value self-enhancement (power and achievement) over openness to change. Those of us who valued openness to change argued that Arnold's inclusion at the reunion would provide both him and us with new opportunities for self-direction and stimulation. For example, because of listening to Arnold's stories, I found I became much more compassionate to the plight of prisoners upon leaving prison. I've begun to donate magazines to prisons and professional clothes to support women interviewing for jobs after leaving prison. I was open to the experiences that Arnold told me about and felt needs I was previously unaware of.

Bring to mind a similar values conflict in your own life and ask yourself whether Schwartz's systems approach to values seems to apply. As you'll likely discover, uncovering the hidden values at work in any interpersonal conflict can help you understand what motivates yourself and others.

Discovering Your Core Values

Values vary in importance. Some values that we hold seem nonnegotiable. I call these cornerstone principles core values. Like all values, core values arise from a combination of our individual character and upbringing, and the values from our culture that we have locked onto and internalized. When we are faced with a values conflict, our core values generally take precedence.

For example, one of my former university colleagues, whom I will call John, held personal success as one of his core values. Although John also valued cooperation, when a situation at the medical school called for him to choose between an action that furthered cooperation among his colleagues and one that advanced his drive for personal success, he abandoned cooperation in favor of individual achievement.

The same hierarchy of values can be seen in my decision to invite cousin Arnold to the family reunion. I hold friendship and the desire to help as core values. Though I also value respecting others, including those who felt uncomfortable about Arnold's past, my core values easily overcame any concerns I had about violating my respect for others who wished not to include Arnold in our family gathering.

One way of discovering your core values is to look back on important decisions you have made to see what values were involved. My close friend Ellen always wanted to be a nun. She entered the convent after attending a small high school on the convent grounds for young women who wanted to become postulants after graduation. Clearly, living a spiritual life was one of Ellen's important values. Yet, after four years in the convent, close to the time when she was due to take her final vows, Ellen met a man in one of her classes off campus with whom she resonated on all levels, including spiritually. Ellen realized that self-direction, a value associated with openness to change, was a core value for her, while the conservative values of tradition, conformity, and security were less important. Although she knew her decision not to take final vows would disappoint her family and the sisters she had come to love in the convent, she could not abandon her core value of choosing her own path.

Married to Michael for many years now, Ellen feels fulfilled in ways she could never have imagined. She prides herself in being able to lead a spiritual life in the context of her family and community, and she is an active participant in her church, the schools her children attend, and various community and service organizations.

Values in Community

In choosing to live as part of a family within the lay community, as opposed to the religious community of a convent, Ellen's story demonstrates an important principle: In order to unlock our full potential, we must be part of a community whose values are harmonious with our own.

Think of the cells of the body. As we have seen, the potential of individual cells manifests itself when the cells are grouped into tissues, organs, and systems that function together to accomplish the same set of goals. Similarly, in the most effective communities, the values of individual members coincide with those of the community, such that the day-to-day actions of its members and leaders reflect a commitment to a shared set of concerns.

Like individuals, communities have values. Some communal values change according to changing circumstances; others are core values that apply no matter what conditions arise. As is true for individuals, some of a community's values may be overt, while others may be hidden or unconscious. When a situation arises that tests the coherence and purpose of a community, its hidden values are often revealed. Think, for instance, of the response of New Yorkers to the terrorist attack of September 11, 2001. Despite the perception of many that New Yorkers are concerned primarily with power, personal achievement, and pleasure, the events of that day revealed that beneath such surface concerns was a hidden commitment to altruism and benevolence. Mutual core community values led people to volunteer their time, energy, and resources—to support each other physically, financially, and emotionally, in spite of differences.

For example, because public transportation was shut down after the attack, a shoe store owner handed out free sneakers to people in dress shoes walking away from the area of the collapsed World Trade Towers. Vendors gave away free food and water to rescue workers from their street kiosks. Months after the event, volunteers still staffed Red Cross victims' relief centers, accompanying victims' families to the site, and helping people who had lost their jobs or homes apply for assistance. As happened in New York, communities with a set of shared values are able to mobilize their members to act as an effective vehicle for the expression of common goals. Shared values power a community, as effectively as electricity powers light.

By contrast, a community with discordant values is often fragmented and immobilized. I recall a values conflict that split the thousand residents of a small New Hampshire town where I once lived. The argument was over whether to limit growth and maintain the town's rural feel, or to allow a developer to chop down trees and clear land to build a shopping center in the center of town, which had no central business district. People who supported the shopping center hoped that it would create jobs; bring a post office, restaurants, and theater to town; and make the town a more interesting place to live. Residents who opposed the shopping center argued that the development would increase crime and destroy the town's residential character. The value systems underlying each of these arguments are familiar to us from Schwartz's study. Meetings at the town hall turned into bitter displays of distrust and accusation.

As a result of this conflict, the developer backed off the plan and approached a similar town, about thirty minutes away. The reaction of people in this town differed significantly. Though there were some residents who opposed the development, citing reasons similar to those voiced in my town, a vocal group of citizens banded together and offered to work with the developer to design the shopping area so that it would enhance, rather than diminish the town's quality of life. These citizens formed committees to work with the developer to determine the size of the shopping area, the kinds of businesses the area would include, and how the design of the buildings might be made to harmonize with the town's existing architectural character.

In three months, they presented a revised plan for the center at a town meeting. Within a year, the completed shopping area offered residents a traditional New England–style town center, with a grocery store and pharmacy, a diner and a dinner restaurant, a theater and several small shops. Ironically, the new town center became a magnet for the people of the surrounding area, including those from the town that rejected the development. Many of that town's teenagers got jobs as clerks in the shops and theater and as servers at the restaurant. Because of the rejecting town's distrust of outsiders and its unwillingness to be open to change, it missed an opportunity to realize many benefits.

Individual members of a community respond to the values held by the community in different ways. Like the cells of the body, we are continually flooded by the influence of circulating values. To be a functioning member of a community, we must lock onto and internalize enough of its shared values to interact smoothly with others. However, unlike the body's cells that respond automatically to the signaling molecules carried by the blood, we can interact consciously with the values circulating in society, and make choices about which values we adopt. Though the residents of the second town were as concerned as the residents of the first town that the new shopping area not destroy their town's character, they chose to work with the idea of development in a creative way, and to be open to the possibility that change might enhance, rather than diminish, what they loved about their town.

Moreover, if we find that the values of our chosen community are at odds with our core values, we can make changes. To bring our own values back into line with the values of the community, we can take one of three actions: We can reevaluate our values and make changes, either temporary or permanent ones. For example, the New York shoe store owner chose to temporarily abandon his core value of making a profit because he saw an overriding value in helping his fellow citizens face a crisis. Or, we can choose to leave the community and align with another, as my friend Ellen did when she chose to leave the convent and become a wife and mother. Or, we can take steps to reshape the values of the community to which we belong, as did the residents of the town who worked with the developer to overcome the objections of residents who feared change.

Before we can do any of these things, however, we must engage in a process of honest self-reflection that aims to help us make conscious the values we have internalized, and those that are held by the communities to which we belong. This process gives us the information we need to make choices about the values we live by and those we wish to foster in our communities.

EXPLORING YOUR VALUES AND THOSE OF YOUR COMMUNITY

Create a quiet place where you feel very comfortable, relaxed, and supported. Take a deep breath in, knowing that your life is nourished by the energy of breath. As you release the out-breath, remind yourself that you are able to release any conflicts and problems that no longer serve you. Continue to breathe deeply in and out until you feel centered and safe. Gently allow your mind to explore the following situations and questions aimed at revealing your core values.

[] Bring to mind an image of your mother and father as you knew them in childhood. Hear their voices as they would characteristically speak to you. What value message do your hear each of them expressing? How did these messages affect your behavior in childhood? Do these messages still affect your behavior? If so, in what ways?

[] Bring to mind the movies, television programs, or books that you would list as your favorites, or ones that influenced you the most. What value messages do they convey? Are these values still influential in your life? If so, in what ways?

[] Bring to mind the images of your closest friends and business associates. How would you describe the values held by these people? In what ways are these values harmonious with yours? In what ways do the values you hold differ?

[] Take a moment to explore whether and how your values have changed over the course of your life. What values that you held in childhood, or as an adolescent, or as a young adult, no longer seem important? What new values have you adopted as a result of your adult experiences?

[] Now ask yourself which values that you hold seem to be core values. Remember that core values are those principles that seem to apply in all situations. Jot down a list of these core values. Recall, if you can, the source of each value. Reflect more deeply on several of the values in your list. Recall a situation in which you demonstrated that this value is part of who you are.

In the next section of this exercise, we explore the values of the communities in which you live. Bring to mind the various communities of which you are part, including your town or neighborhood, your workplace or business associates, your church or spiritual community, and your social or recreational circle. Choose one of these communities to work with in this exercise, knowing that you can go back and answer the same questions about each of your communities, if you choose.

[] Recall a situation that arose within this community that seemed to reveal its essential character. What values did the actions taken by the community reveal? Were these values shared by all members? Were they overt or hidden? Were they consistent with previous value decisions made by the community or did they represent a change?

[] Now ask yourself whether and to what extent the values of this community harmonize with your core values. Have there been occasions when your values conflicted with the values of the community? What choices or changes did you make as a result?

[] What changes might you make to create greater harmony within your community or to better align its values with your own?

Family as Community

No other community seems to elicit as strong a reaction as do our families. The family is a specific, intimate source of internalized values. As happens in the nervous system when neurotransmitters carry information across the synapses between neurons, values are transferred from parents and other relatives to children as a result of close physical and psychological proximity. While it is most prominent in our early life, the influence of our family's values is present throughout our life, even if we have moved far away or ceased regular contact.

As is true in society at large, members may respond to the value messages communicated within a family in different ways. As we have all experienced, children raised in the same values environment may demonstrate considerable variation in their core values. In my father's large Italian family, a core value was honoring parents and elders. This message was communicated by my father and his eleven siblings to their offspring by a system of rewards and punishments. As I was growing up, I witnessed several times the severe consequences visited on children who failed to honor their parents by marrying spouses of whom their parents disapproved. Sometimes such children were disowned by their family. I remember one cousin, in particular, who defied her father in her choice of a husband. Her father pronounced her dead and would not allow her name to be spoken in the house, nor would he allow her to visit or talk to her mother or siblings. No discussion of this edict was allowed, and the father's authority was not questioned.

This value contrasted sharply with those of my mother's Yugoslav family, who valued harmony, although the enforcement of this value often seemed similarly limiting. Many times at holiday dinners, my mother would request that I not express my political views because they did not agree with the family's views. The family's value of harmony at all costs conflicted with my core values of social justice and free expression. Despite my mother's requests, I would often engage in political conversations that ended in discord. As a result of this values conflict, I stopped attending family dinners.

Our strongest emotions frequently rise up in the context of our families. Researchers have used MRIs to measure the activation of emotional structures in the brain when people make difficult moral decisions.[3] The studies suggest that emotional structures in the brain are significantly more active when we make moral decisions that have personal content than when we make impersonal moral judgments.

For example, study participants were asked to imagine a trolley headed for five people who would be killed if the trolley continued on its course. In one scenario, participants were told that they could throw a switch that would divert the trolley to an alternate set of tracks, where it would kill one person instead of five. Most participants chose to throw the switch. In a second scenario involving the same trolley, participants were asked to imagine standing next to a stranger on a footbridge that spanned the tracks between the trolley and the five people likely to be killed. Participants had to decide whether to knock the stranger off the bridge onto the tracks below, which would kill the stranger, whose body would block the trolley's progress, but save the five other people. The majority of participants chose not to knock the stranger off footbridge.

The MRIs of the participants' brains as they considered these two dilemmas showed marked differences. The device is able to measure increases in blood flow in particular regions of the brain. Increased blood flow indicates that neurons in the region are firing more often. The study found that regions in the cortex associated with emotions were significantly more active when personal moral decisions were being made. Participants viewed the decision of whether to push someone to his death as more personal than deciding whether to throw a switch, which would also have resulted in one death.

The different levels of emotional involvement seem to explain why nearly all participants in the study concluded that it was acceptable to sacrifice one life for five in the switch scenario, but not in the footbridge scenario. Further, those rare individuals who judged it appropriate to push someone to his death took longer to reach that decision because their emotions interfered with their process. The study demonstrates that emotions are a functional component of personal moral decisions.

Becoming aware of the impact of our emotions on the process of decision making can help us when we are faced with values conflicts within the family. One useful way to think about these conflicts is to explore the differences between the values of our family of origin—our parents and other relatives—and the values of our adult family—our spouse, partner, or close friends and associates. As adults, we can make choices about how we construct our adult family so as to help make sure that its values are harmonious with our own. However, we should ask ourselves whether and to what extent our emotions are impacting our choices.

In my own case, I wanted my adult family to value partnership and communication. However, my first husband, who happened to be Italian like my father, informed me, while we were both in graduate school, that I would have to delay my studies until he finished his own. He was unable to explain clearly why he had reached this decision. I reacted emotionally to this statement, much as I had reacted to seeing my cousins abandoned by the family when they refused to comply with their father's wishes. My strong emotional response blocked any possible resolution to this dilemma. The conflict marked the beginning of the decline of this marriage, which ended in less than a year.

Initially, my second husband valued our partnership. However, as his businesses met with legal and financial troubles, he experienced depression, often remaining secluded and withdrawn for weeks. As communication between us deteriorated, I felt progressively abandoned. Unable to engage my husband in seeking help, I simply gave up. After seven years, this relationship ended as well.

I've been married to my third husband for twenty-four years. My husband values partnership and communication as greatly as I do. This shared core constitutes a solid foundation for our union. The problems that did arise between us, particularly in our early years together, had clear links to the emotion-laden value conflicts of my childhood. When my husband and I had a serious disagreement, my tendency was to run away from the conflict. In this response, I was mirroring the way I fled my mother's family's dinner table and the way that my father's family abandoned people who did not agree with family edicts. But because I was strongly motivated to make this marriage work, I examined my desire to flee and made the decision to respond to conflicts in a way that more closely reflected my core values. As I shared this internal process with my husband, our mutual commitment to partnership grew, as did our ability to communicate in spite of differences.

In the face of strong emotional pulls, we can choose to perpetuate the values and emotional judgments of our families of origin or take steps to reshape them. The following exercise will help you explore the emotional choices you made in childhood and how you have chosen to perpetuate these values or to reshape them in your adult family.

VALUES AND EMOTION IN YOUR FAMILY

Use the techniques you have learned to create a quiet, comfortable space in which you feel very relaxed. Take in a deep in-breath, acknowledging that your life is nourished by the energy of breath. As you release the out-breath, remind yourself that you can make the decision to release any emotional residue you may have internalized from your family of origin that no longer serves you. Remind yourself that as an adult, you can make choices to shape your family to harmonize with your core values. Now, with great gentleness, allow your mind to explore the following questions:

[] Bring to mind a scene or event from your early life that represents for you the core values of your family of origin. What emotions are being expressed by the participants in this scene or event? Where are you in this scene? What are you feeling?

[] As you recall this scene or event now, what emotions rise for you? How do they differ from your emotions at the time you first experienced them?

[] What core values are revealed by the words and actions of the participants in this scene or event? What core values are revealed by your emotional reaction at the time of this scene or event?

[] How do the core values that you currently hold compare to those you held at that time?

[] In what ways has this core value shaped the choices you have made about your adult family?

[] What role do you play now as regards values in both your family of origin and your adult family? In what ways do you communicate or demonstrate your values to the members of each family?

[] What conflicts do you experience between your core values and the values of your family of origin and your adult family? How do you characteristically respond to these conflicts? What emotions do you experience as a result of these conflicts?

[] What actions might you take to foster greater harmony within your family of origin and your adult family?

Values and the Work Community

Like other communities, effective corporations, civic and religious organizations, and educational institutions act from their core values. Assessing the values of a group and bringing them to consciousness can help transform ineffective organizations into effective ones. Richard Barrett, a corporate consultant and change agent, describes a process of values audits designed to compare the values of employees and the corporations they work. This comparison, he says, is a first step to building visionary companies with a "strong, positive, values-driven culture."[4]

To create a template for these values audits, Barrett related the hierarchy of human needs defined by psychologist Abraham Maslow[5] to the seven levels of consciousness described by psychologists Charles Alexander and Robert Boyer.[6] Maslow's hierarchy posits four types of human needs—physical, emotional, mental, and spiritual. Barrett linked these four to the seven levels of consciousness defined by Alexander and Boyer. He then designed a series of questionnaires and other assessment tools to measure the level of consciousness and determine the values held by both employees and managers of companies and other organizations.

According to Barrett's template, at the first three levels of consciousness, values are driven by self-interest to meet physical and emotional needs. Employees at these levels value security, relationship, and self-esteem, while companies value survival, relationship, and self-esteem. At the fourth level of consciousness, the focus shifts to mental needs. Employees value personal growth and achievement, while corporations value transformation. The higher levels of consciousness, from levels five to seven, are aimed at fulfilling the spiritual need of serving the common good. At these levels, employees value meaning, making a difference, and service, while corporations value organization, community, and society.

Barrett's values-audit process can be used to reveal hidden conflicts between the values of employees and the corporations for which they work. In one part of the audit, employees and managers are asked to list ten ideal values for their organization. One company included in Barrett's process had a thousand employees and five factories. The company's mission statement listed such corporate values as growth, superior products, innovation, creativity, open and honest communication, teamwork, and continuous improvement. However, the values audit revealed a gap between these corporate values and the personal values held by employees and managers.

Both employees and managers listed similar personal values: integrity, respect, truth, honesty, creativity, responsibility, internal motivation, intuition, patience, and making a difference. On Barrett's scale, these values correspond to levels four to seven of consciousness in that they are aimed at fulfilling mental and spiritual needs. But when managers and employees were asked to list the values of the corporation they worked for, both groups listed values corresponding to levels one to three of consciousness, aimed at self-interest: profit, control, tradition, results, diversity, multiple goals, prestige, and competition.

To revitalize this company, Barrett defined three areas for improvement: first, building a culture of trust and collaboration among managers; second, defining a common vision and a set of values shared by managers and employees; and third, using these shared values to foster greater employee involvement and more focus on the needs of customers. Under Barrett's guidance, the managers of the company engaged in a process of rigorous self-assessment to determine the values that best described their individual leadership style. The results of this self-assessment were compared to values assessments of each manager by as many as a dozen colleagues. The information gathered from these measurements was used to devise a leadership development program for each manager. Visionary leaders, Barrett believes, hold the "keys to increased productivity and creativity."[7]

As Barrett's work makes clear, values are pivotal to both corporate success and employee satisfaction. When employees' work no longer feels meaningful to them, they become dissatisfied and corporate morale drops. We have all experienced times when our core values and the values of the company or institution for which we worked did not harmonize. A friend named Claire, who worked as an editor for a small publishing company attached to a nonprofit organization, began to understand that her value of achieving credibility and profitability for her company did not harmonize with the nonprofit's value of promoting its political views. My friend proposed that the board of the nonprofit and the staff of the publishing company meet to craft a mission statement, in the hopes that the process would reveal and help resolve the conflict. Instead, the board of the nonprofit met in private, drafted a mission statement for the publishing company, and delivered it to Claire and her staff. Soon after, Claire's dissatisfaction caused her to leave her position and seek more meaningful employment.

In previous exercises, you've examined the values you internalized from your culture at large and from your family in particular. Now we explore the values of your workplace and the relationship between the organization's values and your own.

EXPLORING VALUES IN YOUR WORK COMMUNITY

In a quiet, reflective state, consider the following questions:

[] How would you describe the core values of your workplace?

[] In what specific ways are these values communicated to employees?

[] In what specific ways are employees rewarded for supporting these values?

Richard Barrett lists several qualities that characterize a visionary organization. Consider whether or how well these qualities apply to your workplace:

[] Does your workplace make a long-lasting commitment to learning and self-renewal?

[] Is it continuously adapting itself based on feedback from internal and external environments?

[] Does it make strategic alliances with internal and external partners, customers, and suppliers?

[] Is it willing to take risks and experiment?

[] Does it have a balanced, values-based approach to measuring performance, including such factors as financial results; efficiency, productivity, and quality; continuous learning and self-development; organizational cohesion and employee fulfillment; and contribution to the local community and society?

Now, ask yourself:

[] What role do you play in regards to values in your workplace?

[] In what ways do you contribute to making your workplace a positive, values-driven culture?

[] What contribution could you make that you are not making now?

New Vibrancy from Old Tensions

When the tensions between individual and collective values are addressed in creative ways, with the focus on shared goals, both individuals and communities benefit. Communities are laboratories in which we can experiment with the effects of shifting perspectives. Because communities at every level include members with diverse values, interacting with others in community gives us the opportunity to try on new ways of being, and to make choices that add to the vibrancy of our individual and collective lives.

I've kept in touch with my cousin Arnold since my family reunion. In struggling to adjust to life after leaving prison, Arnold realized that tension between former prisoners and the rest of society often leads to prisoners committing new crimes and returning to the prison communities in which they have come to feel at home. Arnold decided to try to do something to address these tensions. He approached a middle-size company in his home town that was in need of reliable employees to work the night shift on its production lines. The company's mission statement included such values as community service, innovation, profitability, and continuous improvement. Arnold offered to help the company start a program to recruit and employ released prisoners. He explained that released prisoners had special needs that made them ideally suited for the positions the company was seeking to fill. For instance, former prisoners would particularly value a job they could begin immediately with minimal training, as they were given only one hundred dollars to live on when they left prison. Moreover, many parolees had to meet weekly with counselors and parole officers, which was difficult to do when they were working the day shift. Arnold's rehabilitation program would meet the needs of former prisoners, as well as fulfilling the company's commitment to community service and innovation.

Within three years, thirty-five former prisoners had successfully made the transition to life outside as a result of Arnold's values-based program. Moreover, when the governor of the state recognized the company for its successful work, media attention improved the company's visibility, leading to new clients and higher profits. Values made conscious and acted on within community always lead to new vibrancy and growth.

CHAPTER NOTES

1. Robert C. Nordlie, James D. Foster, and Alex J. Lange, "Regulation of Glucose Production by the Liver," *Annual Review of Nutrition* 19, no. 1 (1999): 379–406.

2. Shalom H. Schwartz, "Are There Universal Aspects in the Structure and Contents of Human Values?" *Journal of Social Issues* 50, no. 4 (1994): 19–45.

3. Joshua D. Greene et al., "An FMRI Investigation of Emotional Engagement in Moral Judgment," *Science* 293, no. 5537 (2001): 2105–2108.

4. Richard Barrett, *Liberating the Corporate Soul* (Boston: Butterworth-Heinemann, 1998).

5. Abraham H. Maslow, *Toward a Psychology of Being* (New York: Van Nostrand Reinhold, 1968).

6. Charles N. Alexander and Robert W. Boyer, "Seven States of Consciousness," *Modern Science and Vedic Science* 2, no. 4 (1989): 325–364.

7. See note 4 above.

CHAPTER 9

witness and use amplification

IMAGINE AN ARRAY OF A THOUSAND DOMINOS, arranged so that when the first domino falls, all of the other dominos fall, one after the other. This cascading effect, in which one action leads to another to produce a large outcome, is called amplification, a principle that operates both in the body and in community. The arrangement of dominos needed to produce this overall effect is the structural requirement for amplification. The falling of the first domino in such a way so as to trigger the sequential fall of all the other dominos in the array is the action requirement for amplification.

In the body, the principle of amplification allows cells and systems to act in concert to produce a larger or more complex outcome than any single cell or system could produce alone. Like the array of dominos, the body's systems are structured to amplify the effect of a triggering signal. This signal, analogous to the fall of the first domino, usually comes from the nervous system or from hormones released by the endocrine system. When a cell or system receives this signal, amplification sets up a repeating pattern within a cell or system or between cells and systems to maximize the body's response.

Let's examine how the principle of amplification operates in a body process with which we are all intimately familiar, the daily cycle of sleeping and waking. The sleep-wake cycle is actually amazingly complex, involving various parts of the brain and nervous system, and several hormones produced by the endocrine system and carried by the blood through the circulatory system. The amplification of the sequential actions of all of these systems makes possible the daily slowing and acceleration of the body's metabolism.

How does this massive change occur day after day? Several research studies have explored the question, What controls the body's sleep rhythm? Subjects participating in these studies were kept in laboratory-controlled, quiet environments devoid of light and cues of time for as long as eighty-nine days.[1] Outside the laboratory, other subjects were isolated in natural caves for as long as 205 days. The diaries of these cave dwellers, as well as the measurements taken of those isolated in laboratory environments, provide clear evidence of a regular sleep-wake cycle. The studies clearly demonstrated that the signal for sleep arises from within the body, rather than from external cues. While people isolated from external cues cycle through sleep and waking on roughly a 24-hour schedule, normal life requires resetting of our body's internal clock to the light/dark cues from the external environment.

Where is the body's internal clock, and how does it operate? Scientists isolated paired groups of about ten thousand neurons, called suprachiasmatic nuclei (SCN neurons), that are located in the hypothalamus region of the brain. Researchers discovered that, like the fall of the first domino, changes in the activity of the SCN neurons trigger a cascade of effects tied to the periodic slowing and acceleration of all the systems in the body.[2]

Let's look in more detail at one such SCN neuron-triggered change critical to inducing sleep. During subjective night—some number of hours after the beginning of day for people isolated from light/dark cues, or the onset of actual darkness for the rest of us—signals from SCN neurons travel to other neurons in the hypothalamus, then down the spinal cord, and out to a part of the autonomic nervous system. These signals stimulate neurons in the autonomic nervous system, whose pathways travel along the arteries, to send a signal to the pineal gland, which lies in the brain between the cerebral hemispheres. This signal stimulates cells in the pineal gland to produce an enzyme that stimulates the synthesis of the hormone melatonin. When it is released, the melatonin activates neurons that contain melatonin receptors, such as the neurons in the brain stem that control the sleep/wake cycle. One to two hours after the level of melatonin rises in the body, sleep ensues. Additional pathways from the SCN to neurons in the brain stem and spinal cord amplify the effects of the firing of SCN neurons on heart rate, breathing, and digestion.

Changes in the levels of neurotransmitters and hormones are just some of the many physiological effects induced by the firing of the SCN neurons. The firing of the SCN neurons fulfills the action requirement of amplification. The neural networks and arteries along which the various effects of this triggering action travel through the body also fulfill the structural requirement of amplification.

In this chapter we explore amplification in the context of community and the changes that a single trigger can bring about, sometimes leading to the transformation of society. We also explore ways that you as an individual can amplify your values or beliefs within your community to trigger positive social changes.

Amplification in Social Action

As happens in systems of the body, groups and individuals in communities amplify signals to bring about far-reaching changes. The civil rights movement is a wonderful example of amplification in action.

On December 1, 1955, Rosa Parks, a forty-three-year-old graduate of Alabama State Teachers College, took the last available seat behind the white people on a segregated bus in Montgomery, Alabama.[3] A few stops later, some white people got onto the bus. The bus driver told Rosa and several other black people to stand up and give up their seats. Three people stood up; Rosa did not. The bus driver told Rosa that he would have her arrested; she told him he could do that. The driver stopped the bus and two policemen got on. One policeman asked her why didn't she stand. Rosa said she didn't think she should have to stand up. The policeman told her that "the law is the law," and he arrested her.

Rosa was unaware that her refusal to stand up would be the trigger that would amplify a movement for change. On December 5, the day Rosa was found guilty, the black people of Montgomery refused to use the buses. The Montgomery Improvement Association was formed and Dr. Martin Luther King Jr. became the group's leader and spokesperson.

That evening a meeting was held at the Holt Street Baptist Church. At that meeting, the Montgomery Improvement Association declared a formal boycott of Montgomery buses. The boycott lasted 382 days and brought Rosa Parks and Dr. Martin King Jr. into national prominence. On November 13, 1956, the United States Supreme Court proclaimed the segregation laws of Alabama illegal.

Sociologists have analyzed how and why actions such as Rosa's were successful in triggering social action. Alden L. Morris, a sociologist from Northwestern University, points to the importance of the civil rights movement for social organizations within local communities, which he terms "local movement centers."[4] These centers mobilize, organize, and coordinate collective action aimed at attaining common ends. The black churches in Montgomery and their charismatic pastors, such as Dr. King, functioned as local movement centers to catalyze social change. They and the organizations they spawned fulfilled the structural requirement for amplification to occur.

The world in which African Americans lived prior to the civil rights movement was a devastating attempt "to impress on Blacks that they were a subordinate population by forcing them to live in a separate inferior society. . . ."[5] As a result, when news of Rosa Park's action reached the black people in Montgomery, the psychological and network structures were in place, and people were ready to act. Belief in the core idea of equal rights amplified the fundamental American values of freedom and equality, which resonated with diverse elements of African American society.[6]

The success of the Montgomery boycott shifted the views of black leaders about what was possible for the civil rights movement. In 1957 black ministers established the Southern Christian Leadership Conference (SCLC), with Dr. Martin Luther King Jr. as its first president. Also that year, Congress passed the Civil Rights Act, which made it a federal crime to interfere with a citizen's right to vote. Further civil rights legislation protected the right of all Americans to vote, expanded desegregation, and expressly forbid discrimination of any kind in housing and employment.

While racial discrimination continues to exist in the United States, the civil rights movement has been a major force establishing nonviolent action as an effective means of social change. The civil rights movement provided the template and impetus for later movements, including the student rights movement, the farm worker's movement, the gay and lesbian rights movement, the women's movement, the Native American rights movement, the environmental movement, and the disability rights movement. Clearly, the impact of Rosa Park's action has amplified beyond Montgomery, beyond Alabama, and even beyond the United States.

The Amplification of a Family's Values

Amplification in community requires supporting structures to relay the triggering signal and to mobilize responses. Among the most important of these supporting structures is the family. Within society, the family functions as a social unit that supports its members in bringing about social change.

A researcher conducted a long-term study of three families who moved to small farms. After moving to their farms, they established new relationships with other families and created social structures, including cooperatives that provided informal and formal opportunities to learn from one another and improve their farming practices. The families were able to establish these new connections because the behavior within each family had taught them how to participate in social groups and establish relationships.[7]

As we saw in chapter 8, the most important learning that takes place within the family is the development and transmission of core values. A family's core values are the signal the family is prepared to send to the larger community. Pivotal events in a family's life, such as the description of the three families who moved to small farms, can trigger the release of that signal. For instance, a core value for the farm families was trust. This value was developed first within each family unit. When the families moved to the farms, that trust was transmitted to others, leading to the establishment of trusting relationships with other families and groups.

The same principles operate in larger social units, as we saw in the beginnings of the civil rights movement in Montgomery. There, the black churches functioned as chosen adult families who shared a set of core values, including beliefs in freedom and equality. The trust built up among the members of these churches as a result of many years of shared struggle and suffering bonded the members into cohesive units, primed to transmit these values to the larger community when they were triggered into action by a pivotal event.

The amplification of a family's values within the larger community can take many forms. It can, for instance, bring about changes within neighborhoods, schools, parks, recreational associations, spiritual organizations, as well as places of business. As is true in all instances of amplification, once transmission of a signal has been triggered, a cascade of effects begins, which can impact life at many levels.

A woman I know, Jennifer, grew up in a family in which financial security was valued highly. Jennifer's mother died when she was still very young, so she spent much time with her father, who taught her about money and investments. Before she was old enough to understand fully what he was saying, her father explained that financial security was not an end in itself. Sound investments, he said, would provide the stable ground on which a family could build its future, and from which its members could draw the means necessary to fulfill their dreams.

Though Jennifer majored in business in college, her brother, Austin, had always been slated to inherit her father's investment firm. After graduation, Jennifer took a job as an analyst at another firm in a nearby town. Then two pivotal events occurred: Austin was killed in an automobile accident, and her father had a severe heart attack that forced him into early retirement. Eager to help out, Jennifer took on servicing her father's clients in addition to her own.

As the months passed, Jennifer began to get the sense that her father was uneasy about her working for his firm. She started hearing rumors that her father might be looking to sell the company. Confused about why her father might object to her assuming leadership of the family business, Jennifer approached her father's long-time friend and associate, Jebb. The story that Jebb told her answered Jennifer's questions.

Several years before Jennifer was born, her parents had had another daughter, Elaine. Because the family business was just getting started, Jennifer's mother had insisted on returning to work soon after Elaine's birth. While Elaine was in the care of a babysitter, the child suffered a seizure. By the time her parents reached the hospital, after a frantic call from the babysitter, Elaine was dead. Jennifer's mother never fully recovered from Elaine's death and died a few years after Jennifer was born. Her father blamed himself for allowing his wife to return to work so quickly. These factors influenced his decision not to allow Jennifer to enter the family business. Secretly, her father feared that working for the firm would somehow cause Jennifer's death as well.

Jennifer thanked Jebb for telling her what her father could not. A whirlwind of thoughts, emotions, and memories swirled in Jennifer's mind. Her father's concern, the basis for his not asking her to join the family firm, did not make her angry, just sad. Jennifer understood how deeply pained her father must have been about the deaths of his infant daughter and then his wife. Now, with her brother Austin's death, she was all that was left of the family. Although she understood her father's desire to protect her, she also felt that it was her responsibility to keep alive the family's values and to transmit them to others.

The next day, Jennifer went to visit her father. Holding his hand, she told him what Jebb had revealed about the family's story. She also told her father that more than ever, she was motivated to use the business as a way of transmitting the values he had taught her to other families. The two talked for a long time. In the end, Jennifer's father asked her to come into his firm and take over its leadership. He knew that her values were sound and wanted them to infuse and shape the business.

Soon after assuming control of the business, Jennifer changed the name of the company to Family Investments. She held a series of meetings in which she communicated her vision for the company to the other members of the firm. With their support, she instituted a series of programs to educate the families of current clients and to invite new families into the program. Because her vision for the company was based on her family's values, Jennifer's firm benefited many people, including families who had never considered themselves able to participate in an investment program. Her innovative ideas helped finance the dreams of many families in her community, including providing funds for college and professional schools, new homes, and new business ventures.

As with the farm families, open boundaries allow values to be transmitted from family groups to the larger community. These open boundaries are evident in Jebb being able to talk to Jennifer about her family's story, in Jennifer being able to talk openly about the past with her father, and in Jennifer being able to communicate her vision for the company to the other members of the firm. With this trust and openness as a basis, Jennifer's firm was successful in amplifying the values held by her family and transmitting them to others in her community.

Social Networks: Pathways for Amplification

Our informal social networks, made up of relatives, friends, business associates, neighbors, and others, function as the structural basis for amplification in community. We might think of these networks as parallel to the neural pathways and vascular highways of the body. Perhaps you've heard the saying, "It's not what you know, but who you know." Underlying this cliché is an important principle tied to amplification: the larger and more complex your social network, the easier it is for your ideas and values to spread to others.

History provides a fascinating example of this principle. In his book *The Tipping Point*,[8] historian Malcolm Gladwell recounts the events of April 18, 1775. On that afternoon, a young boy overheard a British army officer speaking to a colleague in a livery stable in Boston. The officer said something about there being "hell to pay tomorrow." The boy ran to Paul Revere's home to tell the silversmith what he heard. Earlier that same day, Revere had learned of an unusual number of British officers gathered on Boston's Long Wharf. He and his friend Joseph Warren had became convinced that the British were about to march on Lexington and arrest colonial leaders John Hancock and Samuel Adams.

Later that evening, Warren and Revere met and decided to warn the communities surrounding Boston of the British plans. Revere crossed Boston Harbor to Charleston, jumped on his horse, and began his midnight ride. In two hours, he rode fourteen miles between Charleston and Lexington, knocking on doors, and telling people to spread the word to others. The next morning, when the British began their march on Lexington, they met with organized and fierce resistance.

This much of the story is well known. However, Gladwell writes that another revolutionary by the name of William Dawes also made a ride that April night, spreading word of the impending British attack to communities to the west of Boston. Unlike Revere's ride, however, Dawes's ride did not set the countryside afire. What accounts for the difference? Gladwell theorizes that the difference may lie in the number of social connections each rider enjoyed. Revere was a fisherman and a hunter, a card player and theater lover, a frequenter of pubs and a successful businessman, a member of the local Masonic Lodge and several social clubs. He was described as a man with an uncanny genius for being at the center of events. His funeral was attended by "troops of people," as noted in a contemporary newspaper account.[9]

By contrast, according to Gladwell, Dawes was a man with a "normal social circle," which suggests that he did not enjoy as extensive a social network. Gladwell concludes that Dawes simply did not have the social connections that would make people remember him riding through the town that night.

Our social network need not be as extensive is that of Paul Revere to form the structural basis for amplifying social actions. How many people would you guess make up your social network? Who are they? Sociologists interviewed 712 people by telephone from across Florida.[10] Their sample comprised 58 percent women, with an average age of 44. More than half—57 percent—of the respondents were married, 92 percent were white and the average annual income was more than $45,000. These characteristics match the demographics for the state of Florida.

The average social network for those surveyed included 432 people. Of these, 19 percent were family by blood; 11 percent were family by marriage; 19 percent were "step relations," a term coined by the researchers to designate relations through another, such as "my boyfriend's sister" or "my ex-boyfriend's ex-girlfriend"; 18 percent were work-related colleagues; 8 percent were "situational" relations, another term coined by the researchers to include responses such as "because we happen to be in the same place a lot"; 7 percent were school associates; 4 percent were connected by religion; 6 percent were neighbors; and 4 percent were connected through hobbies or organizations.

This representative sample may or may not reflect your social network. Rosa Parks's network, for example, almost certainly exceeded the typical network profiled in the Florida study. Her national associations through the NAACP, as well as her church and black community connections in Montgomery, expanded her network so that her action of refusing to give up her seat sparked a reaction, whereas another person's might not have. Personality may also be also a factor. Gladwell describes Paul Revere as a gregarious and intensely social connector. He was a joiner, one of only two men who belonged to all seven revolutionary groups in Boston. Most of others who were members—more than 80 percent—belonged to just one group.[11]

Do you want to make a difference in the world? The clear message from cellular wisdom is, make connections! Across the world, as throughout the body, networks of connections stand ready to amplify triggering signals.

EXPLORING YOUR FOUNDATIONS FOR SOCIAL ACTION

In a quiet, reflective mode, recall the core values you explored in chapter 8, including the values you internalized from your family of origin and those you hold within your chosen family. Think back over the past year and answer the following questions:

[] What actions have you taken that express your core values? For instance, to what causes have you donated time or money? Have you volunteered to help with school, community, spiritual, or other activities?

[] Who comprises your social network? In what ways do your interests or personality shape your social connections?

[] To what groups or service organizations do you belong? Have you volunteered in an organization that was helpful to you, such as Alcoholics Anonymous, a breast cancer support group, or a divorce group?

[] Do any of these function as a local movement center?

[] What social actions are amplified by your local movement center?

[] What steps might you take to expand your social network?

[] What future actions appeal to you as potential ways to amplify your values?

[] How have you provided your children with a model for serving the needs of others?

Triggers for Amplification

What triggers amplification in community? As we have seen, the action of a person with a well-developed social network can set off a chain reaction of amplified effects. However, what we say, as well as what we do, can trigger social change. Dr. Martin Luther King Jr.'s "I Have a Dream" speech encapsulated the vision of a people and aligned their actions with the greatness of their goal.[12] His words energized the civil rights movement and amplified it beyond the South to the national and even the international level. The speech has been translated into more than twenty-five languages. Even today, when I read his words at workshops, I witness a quickening of spirit. It's as though something in the language itself acts like a triggering signal in people's minds and hearts. Notice what happens inside you when you read these words:

I say to you today, my friends, even though we face the difficulties of today and tomorrow, I still have a dream. It is a dream deeply rooted in the American dream.

I have a dream that one day this nation will rise up and live out the true meaning of its creed: "We hold these truths to be self-evident, that all men are created equal."

I have a dream that one day on the red hills of Georgia, the sons of former slaves and the sons of former slave owners will be able to sit down together at a table of brotherhood.

I have a dream that one day even the state of Mississippi, a state sweltering with the heat of injustice, sweltering with the heat of oppression, will be transformed into an oasis of freedom and justice.

I have a dream that my four little children will one day live in a nation where they will not be judged by the color of their skin but by the content of their character. I have a dream today. . . .

I have a dream that one day "every valley shall be exalted, and every hill and mountain shall be made low; the rough places will be made plain, and the crooked places will be made straight; and the glory of the Lord shall be revealed, and all flesh shall see it together."

This is our hope. This is the faith that I go back to the South with. With this faith we will be able to hew out of the mountain of despair a stone of hope. With this faith we will be able to transform the jangling discords of our nation into a beautiful symphony of brotherhood. With this faith we will be able to work together, to pray together, to struggle together, to go to jail together, to stand up for freedom together, knowing that we will be free one day. This will be the day when all of God's children will be able to sing with a new meaning:

"My country, 'tis of thee, sweet land of liberty, of thee I sing. Land where my fathers died, land of the pilgrim's pride. From every mountainside, let freedom ring!"

And if America is to be a great nation this must become true.

And so let freedom ring from the prodigious hilltops of New Hampshire.

Let freedom ring from the mighty mountains of New York.

Let freedom ring from the heightening Alleghenies of Pennsylvania.

Let freedom ring from the snowcapped Rockies of Colorado.

Let freedom ring from the curvaceous slopes of California.

But not only that: Let freedom ring from Stone Mountain of Georgia.

Let freedom ring from Lookout Mountain of Tennessee.

Let freedom ring from every hill and molehill of Mississippi.

From every mountainside, let freedom ring.

And when this happens, when we allow freedom ring, when we let it ring from every village and every hamlet, from every state and every city, we will be able to speed up that day when all of God's children, black men and white men, Jews and Gentiles, Protestants and Catholics, will be able to join hands and sing in the words of the old Negro spiritual:

"Free at last! Free at last!
Thank God Almighty, we are free at last!"

If we want our words to trigger social action, we must try to understand what it is about King's speech that made it such a powerful catalyst. Leadership experts say a particular combination of vision, personality, language, and style that makes a leader effective in bringing about change. An aspiring leader is able to recognize an unexpressed need in a group that is unhappy but has not done anything to change its circumstances. Seeing this need, the aspiring leader will "typically articulate moral, ideological, or other value implications of that situation for a particular constituency." The leader then "articulates a vision of change that promises them [the group] a better future."[13]

But a charismatic leader does not stop with articulating a vision of the future. Rather, he (or she) challenges his constituency to help bring about change, and expresses strong confidence in their ability to do so. The leader's expressed confidence in the constituency galvanizes them to take action and realize their collective vision.

What can we learn from this analysis? First, to trigger social change, we must be perceptive enough to recognize a social situation that is conducive to change, and to identify a group of people who can be motivated to take action. Second, our words and actions must address the people in this group at the level of values, morality, or ideology they can identify with. As we saw in the case of Jennifer's Family Investments firm, projects for change based on values considerations are easy to grasp and, ultimately, realize. Third, our words must articulate clearly the vision of a better future based on these shared values. And finally, we must express confidence in the ability of people to help bring about needed changes. When these considerations come together, as they did for Dr. King in August 1963, a single speech can topple the first domino and trigger an amplification process that leads to sweeping social changes.

Speeches like Dr. King's may seem irrelevant to our lives or those of our communities, yet that is not the case. Each day, we influence others by the words we say and the actions we take. In order to be effective leaders, whether in our family, our workplace, our neighborhood or community, or on the national or international stage, we can learn much from the example of charismatic leaders like Dr. King. A good place to begin is by asking ourselves, "What is my vision?" King was guided by a vision of equality and freedom. Statements that articulate our personal and community vision encapsulate an energy that reverberates throughout our lives and our communities each time we speak or act in alignment with them. Holding a vision can bring a cohesiveness to our actions and impart meaning and focus to our personal and communal lives.

Author Laurie Jones suggests that we take the time to create a personal mission statement that articulates our values and goals. This statement, Jones writes, can "act as both a harness and a sword—harnessing you to what is true about your life, and cutting away all that is false."[14] If you aspire to become a trigger for amplified action in some sphere of your life, writing a personal mission statement is a wonderful way to begin.

WRITING A PERSONAL MISSION STATEMENT

Make a date with yourself to articulate your mission statement. Choose a time when you can be quiet and undisturbed. Surround yourself with whatever helps you to relax and explore.

[] Bring to mind those activities in which you engage that bring you the most personal satisfaction. Spend some time thinking about why each activity is meaningful. What changes does engaging in each activity bring about in you?

[] Again recalling the core values you articulated in chapter 8, jot down a few phrases that begin, "I do something meaningful for myself when I . . ."

[] After you've written a few phrases, read over your list with an intuitive eye. Which statements jump out as the most significant? Which trigger an emotional reaction or quicken your pulse?

[] Now write a short statement that expresses those significant actions as an aspect of your goal or mission for your life. Keep in mind that your goal or mission should express a vision for a better personal future based on your values, and spark confidence in your ability to achieve that future. For example, you might write:

My goal is to inspire the love of art in children.

My goal to create an environment that promotes personal growth.

My goal is to help people take on responsibility for maintaining their health.

Now bring to mind those activities you engage in as a member of a group that bring you the most satisfaction. Spend some time thinking about why each activity is meaningful to you and to the group.

Jot down a few phrases that begin, "I do something meaningful for and with others when I . . ." Read over your list and assess which activities trigger the strongest response.

Add to or revise your mission statement to include your social goals. As you write and revise, aim to come up with a single statement that expresses your personal and communal goals.

Laurie Jones gives the following criteria for a mission statement: It should be no more than a single sentence long, easily understood by a twelve-year-old, and able to be recited by memory at gunpoint.[10] She gives the following example of a mission statement developed by a labor relations expert: "To uphold, discover, and support trust, honesty, and integrity in all relationships." A woman working at a cancer cure center developed this mission statement: "To inhale every sunrise and look under every rock for the joy life has to offer." The mission statement I developed for my coaching business is, "To freshly and expansively unfold my decision to love and help people evoke their greatness." Each time I voice my mission statement in my own mind, or share it with clients—which I do several times a day—my heart quickens and an energy flows, sustaining and inspiring me. I have no doubt that my mission statement amplifies my ability to trigger change.

Finding Your Role

In our discussion of the sleep/wake cycle, we saw that the suprachiasmatic nuclei (SCN neurons) function as the leaders, signaling or triggering the day-night alterations in many systems. However, without the responding systems, there would be no amplification of those signals and, therefore, no day-night rhythm necessary to maintain body functioning.

The same need for leaders and followers exists in successful social movements. Leaders articulate the vision and infuse people with the energy for its manifestation; they are the triggers for amplification. Followers provide the structural complement critical to amplifying the vision and implementing its goals.[15] In their own way, both leaders and followers serve the central purpose, or mission, of a movement or organization.

A successful social movement also needs supporters. Supporters are people who have no formal association with a movement or organization, but who believe in its mission and empathize with its leaders and followers. The energy infused into social movements by supporters can be as critical as the actions of its leaders and followers. Supporters provide the environment in which a social movement can flourish. Without the support of many people who never participated actively in marches, sit-ins, or other concrete actions, the civil rights movement would never have succeeded. Moreover, it is possible for us to support a variety of social movements, more than we could possibly participate in as leaders or followers.

At some time and in some aspect of our lives, we've all been leaders, followers, and supporters. Finding our appropriate role and acting within it comfortably provides the ideal conduit for our service to the causes in which we believe. Such service contributes to our ability to achieve happiness and well-being through our social interactions. It's important, however, that we not make rigid assumptions about our appropriate role. As our life circumstances change, our role may also evolve, as the following story illustrates.

No one in Mel's company would have characterized him as a leader. For ten years, he had enjoyed his work as a graphics artist with a small advertising agency. Then his company was bought by a larger agency, one of the biggest in New York City. Within weeks, a spokesperson announced that the firm would be moving from Boston to New York. Anyone not willing to make the move would be let go. The announcement was met with dismay by many people in the company. Fred, for example, had an autistic child who was attending one of the best schools in the world for autistic children, located just outside Boston. Fred's family simply could not move to New York. And Fred was not the only talented graphic artist who would be lost to the firm.

As Mel thought about how unfair the situation was for Fred and the others, he grew more and more disturbed. Though he had always been a quiet, introspective person, he began to speak to others in the company about their reactions to the move. As he did, he heard several stories as compelling as Fred's. Mel decided to do something to make the company executives aware of these people's situations.

First, he talked to his supervisor, but received no encouragement. "Be glad you have a job," his supervisor told him. "This is the way things are. What did you expect, anyway?"

The question churned in Mel's mind. What did he expect? Slowly, the answer began to emerge. He expected the company to treat its employees with the same high values they applied to business decisions. Mel had always been proud that his company refused to accept contracts for products with dubious social value, such as cigarettes. The supervisor's question triggered others in Mel's mind. What if "the way things are" wasn't the way they had to be? And, what might he do to help others and to help his company at the same time?

Then Mel had an idea. He recalled reading that telecommuting actually saves money for firms facing a move. He asked the company's chief accountant, a close friend, to help him compare the cost of relocating people who could not move to New York with the cost of providing them the technology to work from home offices in Boston. Armed with data that showed it would be cost-effective for employees to continue working in Boston, Mel flew to New York to meet with the president of the new firm. Mel presented his idea as both a socially responsible action and a cost-saving plan for the merger. After an accountant's review verified the numbers in Mel's presentation, the president agreed to offer the telecommuting option to those employees unable to leave the Boston area. Fred and the others were delighted with the willingness of the new executives to treat them as people with obligations that were incompatible with the move. The firm's flexibility and social responsibility created a basis for strong loyalty to the new company. Mel was amazed at the success of his activities. He experienced a sense of fulfillment and well-being that was invigorating as he continued his creative work.

Mel exemplifies the transition from follower to leader. His empathy triggered a process that led to a new plan that benefited his coworkers. The positive responses of the chief accountant and the president provided the structural basis for amplifying Mel's idea. His sense of fulfillment and well-being issued from expressing his talent for leadership in a particular situation. Nonetheless, once the new plan was implemented, Mel returned to his role as follower, happy to telecommute from his home office in Boston.

What roles have you played in amplifying change in the communities to which you belong? Are there opportunities for you to play a different role? The following exercise can help you find out.

FINDING YOUR ROLE IN SOCIAL ACTION

Settle into your restful place, choosing a time during which you can reflect without being disturbed. Take three or four deep breaths. As you exhale, release any tensions, first physical and then psychological, by seeing them as clouds moving across the horizon away from you, issuing in a clear, bright, internal landscape. When you reach a state of serenity, consider the following:

[] In what situations have you acted as a leader, follower, or supporter in your family, workplace, community, or other groups in your adult life?

[] Choose several of these situations that you enjoyed more than the others.

[] What did you enjoy about the role you played? Be as specific as you can in recalling the components that resonated with your personality or ideas.

[] Does this analysis reveal a particular role that is your most natural way of relating to others? Or do you play different roles depending on the group or situation?

[] After completing this analysis, put it aside mentally, and relax more deeply. Ask yourself, What situation exists in some sphere of my communal life that I would really like to change?

[] Articulate what disturbs you about the situation. Take your time. Recall instances when your response was "That's not fair!" or "That should not happen to anyone!"

[] Now begin to generate *What if?* questions in relation to this issue.

[] Choose one *What if?* question to explore in more detail. What conditions would be necessary to change the way things are to the way things ought to be?

[] What role could you play in helping to manifest this outcome? It's okay if you wish to ruminate over this question for several days.

[] When you feel ready, identify a supporter. Describe to the supporter you've chosen what kind of support you are looking for: A sounding

board? Psychological support? Another's belief in your proposed
action?

[] Act! Take a small, but public step. Call someone, let someone know
you are interested, sign up for something, commit to someone, join a
group you want to work with, support someone who could use help
to continue their social action.

You're on your way. Whatever action you take will provide an opening
for the process of amplification to occur. Step-by-step action forms the
basis for strong and positive action.

CHAPTER NOTES

1. P. Lavie, "Sleep-Wake as a Biological Rhythm," *Annual Review of Psychology* 52, no. 1 (2001): 277–303.

2. Robert Y. Moore MD, "Circadian Rhythms: Basic Neurobiology and Clinical Applications," *Annual Review of Medicine* 48, no. 1 (1997): 253–266.

3. Rosa Parks, *Quiet Strength* (Grand Rapids, MI: Zondervan Publishing House, 1994).

4. Aldon D. Morris, "A Retrospective on the Civil Rights Movement: Political and Intellectual Landmarks," *Annual Review of Sociology* 25, no. 1 (1999): 517–539.

5. See note 4 above.

6. Margaret M. Bubolz, "Family As Source, User, and Builder of Social Capital," *Journal of Socio-Economics* 30, no. 2 (2001): 129–131.

7. See note 6 above.

8. Malcolm Gladwell, *The Tipping Point* (Boston: Little, Brown and Company, 2000).

9. See note 8 above.

10. C. McCarty et al., "Eliciting Representative Samples of Personal Networks," *Social Networks* 19, no. 4 (1997): 303–323.

11. See note 8 above.

12. Martin Luther King Jr., "Address at March on Washington," Martin Luther King Jr. Papers Project Speeches, http://www.stanford.edu/group/King (accessed September 22, 2003).

13. Chanoch Jacobsen and Robert J. House, "Dynamics of Charismatic Leadership: A Process Theory, Simulation Model, and Tests," *The Leadership Quarterly* 12, no. 1 (2001): 75–112.

14. Laurie B. Jones, *The Path Creating Your Mission Statement for Work and for Life* (New York: Hyperion, 1996).

15. Ira Chaleff, *The Courageous Follower* (San Francisco: Berrett-Koehler Publishers, 1998).

CHAPTER 10

give and respond to feedback

LIVING AND WORKING WITH OTHERS—in a family, neighborhood, business, or organization—requires skill in giving and responding to feedback. Almost all human social interactions necessitate that we communicate information clearly to others. Our ability to communicate clear messages about what we think and how we feel is a critical social skill, whether we are giving feedback as a parent to a child running into the street, or as a manager to an employee who is habitually late for work, or as a member of an organization to a proposal for joint action. At the same time, we are constantly receiving messages from others, through what people around us say and do, and through more subtle and indirect signals. Sometimes, circumstances require that we respond directly to feedback, as when our partner expresses unhappiness about something we have said or done, or when we receive a critical evaluation for our on-the-job performance. Other times, the appropriate response is internal and requires more a shift in attitude than a change in outward behavior.

As with the other life skills we have been exploring in this book, our body provides a wonderful model for how feedback is best given and responded to. Let's look, for example, at how the body controls the level of glucose it gets from the food we eat. As you probably know, glucose is a major nutrient that provides energy to power the body's processes. Because the level of glucose in the body is constantly changing, the level of this cellular fuel circulating throughout the body must be continuously monitored and tightly controlled to maintain optimum concentrations. The pancreas and liver play key roles in promoting the utilization of glucose and balancing the amount in the bloodstream.[1] After a meal, as glucose levels in the blood rise, specific cells in the pancreas detect the change and respond by secreting the hormone insulin. Insulin promotes the uptake of glucose by muscle and fat cells. Insulin also signals the liver to store excess glucose as a condensed polymer, glycogen, for future use. When additional glucose is needed, stores of glycogen are reconverted back into glucose, restoring the level of glucose in the blood.

This feedback system includes numerous intricate regulating processes utilizing a host of enzymes and transporter molecules. In broad overview, the processes that shape this feedback require three distinct steps: (1) the existing levels of glucose in the bloodstream must be continually monitored; (2) the information must be communicated accurately to relevant cells and organs; and (3) adjustments must be made to bring conditions into appropriate balance. At each step, accurate feedback is essential, for if the level of glucose remains too high, diabetes results, with all of its potential complications; if the level is too low, hypoglycemia results, leading to dysfunctions in many systems, including the brain. Perpetually alert, the body senses, communicates, and responds by sending vital feedback to appropriate sites to sustain health, moment-by-moment, throughout our lives.

Following the body's lead, our ability to give clear feedback to others, take in the feedback we are constantly receiving, and respond in appropriate ways to that feedback has three necessary components. First, we must be alert to, and aware of, the messages others are sending, and we must be prepared to send clear messages of our own. As happens with systems in the body, we must be attuned to monitoring dynamic changes in present conditions and respond from that place of awareness. Second, we must hone our ability to hear and communicate social messages. Finally, we must find skillful ways to make adjustments, both in how we send messages and in how we respond to incoming signals from others. In this chapter, we explore each of these areas, with the understanding that learning to give and respond to feedback skillfully is essential to the well-being of our family, organization, business, and community, as well as to our own personal well-being.

Staying Alert to Feedback

One of the best ways we can maximize our ability to stay alert to feedback is by increasing our self-awareness about our characteristic style of interacting with others. Psychologists point out that people display significant preferences in the way they perceive the world and take in information. Some of us are naturally inclined to gather needed information from sensory clues in our external environment. We are tuned to be alert to sights and sounds, smells and tastes; to what other people say and do; and to the rich details of our immediate surroundings. Personality theorists call people with this orientation sensing types.

Others of us are far better at gathering information from internal sources. If this is our natural bent, we are more alert to overall impressions and feelings, to unseen possibilities, and to the inner world of the imagination. Psychologists call people with this orientation intuitive types.

Research indicates that business managers, supervisors, and executives, as well as people who work in retail, banking, telephone, and accounting businesses, have a preference for the sensing process.[2] Not surprisingly, people who work as counselors, educators, social workers, and rehabilitation therapists tend to favor the intuitive process. Interestingly, in the current business climate, the occupational distinction between these two personality types seems to be blurring: previously undervalued by executives oriented to bottom-line issues, intuitive types are now increasingly sought by businesses because they bring adaptability and people-orientation into the decision-making process.

It turns out that an awareness of one's natural tendency toward sensory or intuitive information gathering can help both types interact more meaningfully in the workplace. People of opposite orientations can sometimes talk past one another—the sensory type alert for overt clues in speech and action, the intuitive type focused on an overall impression of the interaction. But people with dissimilar orientations can interact meaningfully. They need to make the effort to recognize that people have different preferences, and develop strategies to relate better with people who take in information in a different way than they do.

For instance, sensory types, have a tendency to engage in communication on a need-to-know basis, believing that other communication is a waste of time. They are objective, analytical, and quick decision makers; brief and businesslike; and they believe that toughness and power are important. A manager with these personality tendencies would benefit from stretching in the opposite direction when giving feedback to an employee of the intuitive type. The manager might establish better rapport by listening rather than speaking, asking rather than telling, and allowing extra time for the feedback session.

Psychologists also differentiate between people who have a learning orientation and those who have a performance orientation.[3] People with a learning orientation seek to develop their abilities and skills and to master new situations. They tend to view feedback in a favorable light as an aid to further development. People with a performance orientation seek to demonstrate and validate their abilities. They may be uncomfortable with feedback and avoid situations in which feedback is given, as the negative judgments often challenge their need to have their performance validated. Moreover, both learning types and performance types react more favorably to feedback that is given in an informational manner rather than in a controlling manner. Informational feedback is perceived as supportive and constructive, while controlling feedback, which stresses demands to obtain specific outcomes, is often interpreted as inhibiting and restraining.

We can see how these various parameters affect our ability to function effectively in the workplace in the story of a man I will call Gerald. When Gerald became the executive director of a nonprofit dance company in a mid-size Southern city, his future seemed unlimited. He was charismatic, visionary, and brilliant. His powerful presence, combined with an extraordinary command of language, announced that he was a person to be reckoned with. One had only to see Gerald in action once, such as the time he used his powerful body language and a few well-chosen sentences to take the company's technical director to task for a missed lighting cue during a performance, to know that crossing him was a bad idea. Word spread quickly among the creative staff, dancers, and volunteers: If you know what's good for you, you'll stay out of Gerald's way and tell him what he wants to hear.

Gerald had big plans for growing the company. The city needed a new performance space for dance and theater events, so Gerald launched an ambitious campaign to raise funds and build a municipal performing arts center. Excited by his energy and vision, local business people and the city administrators believed he could work miracles. He engaged an architect, planned and carried out an expensive public relations campaign, and took steps to secure a site at a time when no one else was building. Though the company's board and fiscal officers were concerned about the mounting expenditures Gerald's activities were incurring, no one pointed out the risk. Accurate monitoring of the fiscal situation slipped, and the accounting process became perfunctory. It was the president of the local bank, which had been backing Gerald's campaign with a line of credit, who finally called a halt to the proceedings. The project was two million dollars in deficit, and the performing arts center was still on the drawing board. Though various solutions were explored for resolving the problem, no strategy was effective. Gerald was asked to resign, and the dance company ceased operation.

How do the personality types we have been looking at help us to understand what went wrong? First, it's clear that Gerald was an intuitive, rather than a sensing type personality. Visionary people are generally intuitives. For them, the world of the imagination can seem as real and as believable as external reality. Gerald's personal charisma allowed him to communicate his vision clearly to others, but his unwillingness to invite feedback from sensing types like the company's fiscal officers meant that there was no reality-based corrective to his grand scheme. Moreover, as a former dancer whose career had been a series of critical triumphs, Gerald was orientated toward performance rather than learning. That's why he avoided situations in which he might hear negative judgments about his plans. Finally, as a manager, Gerald was controlling rather than informational in the feedback he gave to others. By using his personal power and communication skills to demand high standards of performance from the people in the company, he created an atmosphere in which people felt inhibited about voicing their doubts and concerns.

Leaders with strong egos and a passionate vision are easily self-deceived. They may be so invested in making their mark that they cannot let in information telling them it cannot be made in this way, at this time.[4] A courageous follower is someone who takes on the responsibility to minimize the self-deception and find ways of revealing reality to the leader. Gerald's out-of-control behavior, like rising levels of glucose in the body, initiated a fatally dysfunctional state in his dance company. One accountant giving clear feedback at the appropriate time might have saved the situation before it was too late, but sadly, no one was courageous enough to do so.

Spend a few moments thinking about the organizations, businesses, and other social structures of which you are a part. Are you a leader or a follower in these settings? Are there situations within any of these groups that would benefit from your being able to give or receive clear feedback? What steps might you take to become a more responsive leader or a more courageous follower?

Sending and Receiving Accurate Feedback

In a healthy body, multiple systems are continuously sending and receiving feedback. We looked briefly at the feedback mechanisms for regulating the level of glucose in the bloodstream, but many other systems in the body also rely on feedback. For example, sensors on special muscle fibers continually send feedback about the state of tension in the muscles. Even as we stand upright in place, adjustments are being made in response to feedback to maintain muscle tone, allowing us to resist the pull of gravity. As we begin to move, say, by riding a bicycle, various blood vessels dilate and constrict to maintain adequate blood flow to the muscles and other organs, depending on how vigorously we peddle. As our level of exertion increases, our rate of breathing also increases so that we are able to supply the maximum amount of oxygen to our tissues.

These critical adjustments generally do not require our intervention. As we will see, however, people who are working to overcome a problem such as stuttering can learn to become aware of some aspects of body functioning, and adjust them consciously. When they do, they are engaging in one of three main types of feedback: self-monitoring. Two other types of feedback are common in social situations: formal feedback, such as performance evaluations from a work supervisor, and one-on-one feedback, such as advice from a good friend, counselor, or life coach. Each type of feedback provides a different kind of information. All three can be quite useful as long as the feedback meets three criteria for effectiveness: (1) changing conditions are monitored moment-by-moment; (2) information about these changes is communicated clearly to the appropriate person; and (3) the person receiving the feedback has the skills and the opportunity to make necessary adjustments.

Self-monitoring. If you have ever had a problem with stuttering or know someone who has, you know that stuttering can be debilitating. People who stutter experience an emotional upheaval that often manifests in stress reactions, negative attitudes, life-style adjustments, and a perceived loss of control.[5] Among the most effective treatments for this condition, is a computer-assisted feedback program that gives people immediate and continuous feedback on a number of body parameters. Working with the program allows people to break down the problem into a number of manageable parts and to focus on each one independently.[6]

A number of body functions have been identified as being related to stuttering. These include whether the diaphragm is used during breathing, whether the breath flow is continuous, whether the person exhales before speaking, whether and how much the sound volume rises as the person begins to speak, whether the first sound spoken is prolonged, and whether phonation and phrasing are continuous. A person using the feedback training program is wired to monitor these various functions. Immediate feedback on each of these parameters is displayed on a computer screen so that the person can assess his or her performance. For example, a purple line on the computer screen traces the breathing cycle. Trainees learn to control the line on the screen to help them establish a smooth breath cycle. Similar feedback helps them work with each target behavior in turn, mastering one behavior before moving on to the next.

The negative emotional consequences that often accompany stuttering are addressed in an accompanying program. This training allows a person to take personal responsibility for whether or not stuttering occurs, and to generate improved feelings of well-being, self-esteem, and resilience. Working with a speech clinician, clients respond to a series of questions about their awareness of their speech problem, their self-esteem, their sense of control, and their general well-being. Clients are taught to monitor their own progress using self-administered inventories that measure their fear and anxiety about speaking to others, their assertiveness, their confidence in being able to speak fluently as a result of the training, and their attitudes and feelings about stuttering. Clients record their coping strategies, problem-solving techniques, and self-responsibility goals in diaries. Nondirective counseling helps clients discover their own answers and leads to increased self-responsibility.

This two-part training program has proved to be very effective. Adults who had stuttered since childhood reduced the incidence of stuttering to a minimal level within a few weeks. Moreover, positive changes to their ability to communicate enhanced feelings of self-efficacy and led to measurable changes in how people perceived themselves and their ability to communicate. Follow-up evaluations at six months and one year proved the sustained efficacy of the training program.

The effectiveness of this program demonstrates well the components of all effective feedback: (1) trainees were able to monitor various parameters of their speech moment-by-moment; (2) the information was accurately and clearly communicated to the person using the program; and (3) trainees were taught to make adjustments necessary and to feel in control of their own process.

In addition to external feedback, such as that provided by speech, subtle cues reflecting our internal states are also accessible as sources of feedback. Psychologically minded individuals are able to experience their internal environment, including thoughts and feelings, in relation to ongoing external events.[7] In addition to its self-monitoring function, psychological mindedness is a social skill through which people gain entrance to the internal states of others by way of empathy.

The ability to empathize with others begins in childhood.[8] Developmental psychologists have found that children as young as two are able to imagine themselves in another person's situation. As adults, individuals vary in their ability to share the experiences of others. Some people are highly empathic; others, less so. Researchers have discovered that empathy for others is linked to emotional intelligence, the ability to perceive, understand, and manage emotions.[9] Such self-monitoring not only helps people manage their own emotions, it also helps them connect empathically with others by inferring their internal states.

Formal feedback. As we saw in the case of Gerald, a lack of feedback can lead to disastrous results. While signs of the dance company's declining financial base may have been apparent to its accountants, no one was brave enough to communicate the unpleasant news to the dance company's charismatic leader. By keeping quiet, members of the company violated the second component of feedback: communicating accurate information to the appropriate people.

Communication alone, however, may not have been sufficient to change Gerald Johnson's behavior. A typical management feedback tool, called 360-degree analysis, is used by an estimated one-quarter of the companies in the United States.[10] The 360-degree feedback process elicits anonymous evaluation comments about an employee from direct reports, peers, and managers with the aim of providing multiple perspectives on an individual's on-the-job performance. The tool is designed with the assumption that negative feedback, defined as ratings from others that are lower than the employee's self-ratings, will create awareness and motivate individuals to change behaviors.

To assess the tool's effectiveness, business experts studied the reactions of 125 adult students enrolled in a graduate-level MBA program to a questionnaire that used the 360-degree feedback model to assess their leadership behaviors.[11] Before entering the MBA program, participants in the study had worked an average of five years, including at least two years as managers. The participants' former managers, peers, and direct reports gave them feedback on twenty leadership skills, including their ability to listen to others, manage disagreements, and foster teamwork. External feedback was compared to the participants' self-ratings on these same skills, and the results were illustrated on a graph.

Formal feedback was presented to the students in a group session led by a facilitator, who explained how to interpret the information on the graphs, emphasized the importance of feedback to personal development, and encouraged participants to schedule one-on-one meetings to explore how to use the feedback for developmental purposes. Specifically, the facilitator suggested that the students could use the information from the study to personalize their MBA programs for maximum individual development. At the end of this group session, the students were asked to complete a questionnaire to assess their attitudes and reactions to the feedback. Students filled out a second questionnaire following the voluntary individual session with the facilitator, during which each student had the opportunity to discuss a personal plan for development. Of the 125 participants in the study, 103 attended this second session. The facilitator rated the students on their attitudes and reactions to the feedback.

The results of these assessments showed that negative ratings, defined as those that were low or lower than the students expected, were not seen as accurate or useful by the students. Negative feedback, the researchers concluded, "did not result in enlightenment or awareness but rather in negative reactions such as anger and discouragement." Interestingly, high ratings were not associated with positive reactions but "merely [with] the absence of negative reactions." Furthermore, while high ratings from managers or direct reports were viewed as accurate, those from peers were not. Maybe that's why feedback from direct reports did not influence the students' reactions as much as feedback from managers and peers did. The study suggested that those who needed feedback the most because they were performing poorly or because they overrated their performance, found it the least useful!

The most hopeful finding in the study was that the students' perception of the usefulness of the feedback increased after the one-to-one sessions. Apparently, follow-up activities helped recipients to deal with negative reactions and work with them constructively. A facilitator can guide recipients by helping them to feel good about their high ratings, and minimize some of the negative reactions to low ratings and suggested developmental needs. The facilitator may also motivate recipients to continue successful behaviors. In other words, the 360-degree analysis tool is most effective if it includes the facilitation follow up with participants.[12]

One-on-one feedback. Let's look more closely at how and why one-on-one feedback is so effective by considering the experience of one of my coaching clients.

Elizabeth, a new MBA, was completing her first year on the job at the company she'd always dreamed of working for. The computer chip manufacturer had been named one of the hundred most socially conscious companies in the country. In addition to on-site day care for employees' kids, generous family leave and flex-time policies, and a state-of-the-art employee health club, the company encouraged employees to volunteer in the community-based project it sponsored to educate adult illiterates. Elizabeth had spent the summer volunteering at the project, teaching reading and office skills, and helping place program graduates in jobs.

Confident she had done well as the supervisor of a hardware development group, Elizabeth was not at all hesitant to participate in the annual 360-degree analysis tool the company used to provide feedback to employees. But she was surprised and dismayed at the results. She called me from her office, very upset, and asked to meet with me later that day. I had been coaching Elizabeth since the early days of her MBA program, so she knew I would be a strong support for her.

In our meeting, we read through the feedback results together and noted those responses that upset Elizabeth the most. She seemed particularly distressed that the feedback from her peers and direct reports described her behavior as controlling. She'd never imagined that to be true about her management style. "I feel personally responsible for the quality of work of each member of my team," she told me. "How can I make sure things are being done right without looking over people's shoulders?"

I suggested that Elizabeth read the feedback as though it were describing the performance of a woman on her staff and imagine what she might suggest to her.

"I'd ask the woman to envision a circumstance that would allow someone to describe her behavior as controlling," Elizabeth told me. "Then I'd suggest that she come up with several scenarios in which she acted differently."

"Let's do that now about you," I suggested. With Elizabeth directing the action, I took the part of a member of her staff as we engaged in role-playing several scenarios in which Elizabeth practiced the new, less controlling behaviors she envisioned.

Finally, I asked Elizabeth if she had thought about who she wanted to be in the company. "The vision of that future self can be a wonderful guide," I told her. "Rather than waiting to become that wise person, simply take steps now to act in ways your future self might, even if it feels sometimes as if you are pretending. In fact, you are pretending, but that's the way all learning takes place."

"I can't believe I had the answers all along," she told me as she was leaving. The next day, when she met with her supervisor to discuss the feedback, Elizabeth was clear and confident. She spoke candidly about her initial reaction to the feedback and about how it had helped her to come to a clearer understanding of how she could improve her management style. By the end of the meeting, she had outlined a plan for changing how she managed her group, with the aim of empowering her staff to take more personal responsibility for quality control and problem solving. Elizabeth's supervisor approved the plan and was enthusiastic about the strategies she had outlined.

Elizabeth's experience in working through her performance review with the help of a professional coach is not unusual. Employees who receive negative feedback often begin by feeling like a victim or by angrily refuting comments that challenge their self-perceptions. Talking about these responses helps them remove layers of self-recrimination and self-doubt, until they begin to see that something has prevented them from acting from their authentic or core self on the job. Some people realize that they've been hiding aspects of themselves for fear of criticism. Once they begin to see who they want to be in their company, they find it easier to strategize ways to overcome the obstacles. The desire to reclaim or realize one's core truth constitutes a powerful drive and empowers people to use feedback to further their development. Some call this approach the wisdom perspective, a way of looking at the self that has recently become the subject of psychological research.[13]

I, too, try to keep the vision of my true self in focus. For each area of my life, I've created an index card with inspiring images that depict my vision of that aspect of my life on one side, and a written description of how I want to behave to reflect that vision on the back. For example, one side of a card displays the words *Exercise* and *Nutrition* and the image of eight people in a rowboat rowing. These statements appear on other side of the card: "My exercising and nutrition are expressions of love for myself. I exercise and eat healthily as an expansive expression of love. My exercise and nutrition are a decision for happiness. My exercise and nutrition unfold as I love moment-by-moment." I keep the laminated cards by my bedside. Each morning, I choose one card as a focus for the day. Holding the vision in mind and evaluating my behavior in response to it helps me to fulfill the first requirement of effective feedback: moment-to-moment awareness. Holding your true self in focus gives you a model to reference as you move toward leading a more fulfilling and purposeful life as a member of your community.

Working with a counselor, therapist, or professional coach can also help you see your daily ups and downs in the context of achieving your ultimate goal. When you are feeling discouraged, a coach can help you remember that wise people always benefit from the mistakes they've made; they use them to refine their vision and move toward it. Legendary college football coach Bear Bryant of the University of Alabama veered from tradition and showed his team movies of their successes, practice after practice, rather than showing them their mistakes. Perhaps that's why the Crimson Tide had an unprecedented record of victories under Coach Bryant.

You can use some of these same techniques to soften your response to negative feedback, and fuel your progress toward being able to act as your most authentic self in relationship to your family, company, and community.

RESHAPING NEGATIVE FEEDBACK

In comfortable surroundings, create an environment that is soothing and relaxing, where you will not be disturbed. Have plenty of water available.

[] Bring to mind a negative comment or piece of feedback regarding your behavior in a community whose values are closely aligned with yours. Without trying to escape or run away from any uncomfortable feelings, allow the emotional sting to rise and then consciously let it go. Imagine that the negative feelings are like smoke that has been blown away by a gust of wind, leaving the air clear.

[] Now imagine that the feedback you received was directed instead at a close friend who has come to you for advice. What words of comfort and support would you offer your friend? How would you advise your friend to feel about the feedback? Remembering that the last thing you want to do is to hurt your friend's feelings, describe in gentle, but specific terms what behavioral changes your friend might make in response to this feedback.

[] Now bring to mind an image of yourself as you would most like to be in relation to the community in question. Make the picture as vivid as possible. What are you wearing? What are you doing? How do others respond?

[] Recall times in your life when you behaved most like this imagined self, regardless of where it occurred. Bring to mind the specific details and pay attention to the positive feelings that arise.

[] Now bring to mind again the incidence of negative feedback with which you began this exercise. Ask yourself what you might do to create positive feelings, such as those you experienced in the past when you were behaving as your most authentic self. Imagine the new scenario in detail, role-playing the new you in your mind.

[] End this process by writing a brief statement that describes your vision of your future self and three specific steps you might take to move toward it.

When you have completed your self-process, consider whether you wish to work on your issue with an outside facilitator. As we have learned, people often make the most progress when they seek advice from a close friend, counselor, therapist, work associate, or professional coach.

Making Adjustments in Response to Feedback

As you may have discovered while working through the previous exercise, the most difficult step in making adjustments in response to feedback is translating your vision of your ideal self into a series of practical steps you are willing to take. In working with my coaching clients, I have found several techniques that aid this process. First, rather than trying to make changes to every aspect of your behavior, identify several specific aspects of the feedback on which you wish to focus. Under the broad category of being too controlling, for example, Elizabeth chose to focus on comments that indicated that she does not allow time for others to process new information, is not flexible about considering alternate solutions, and is excessively goal-oriented.

Next, compare these behaviors to the values you identified in chapter 8 as being your core values, and see what you discover. In considering this step, Elizabeth realized that the personal value she placed on high achievement and reaching her goals was conflicting with another of her core values, respecting the opinions of others. As we discussed this issue, Elizabeth realized that she could bring these values into harmony by involving the members of her team in a cooperative process of goal-setting and problem-solving. In comparing her on-the-job performance to her core values, Elizabeth was engaging second step in her plan for implementing successful changes: determine the value of making a change. Following this step, she would envision the outcome of changes to be made over the next six months, determine a series of action steps to be implemented within the next thirty days, carry out these steps and evaluate their results, and develop strategies for maintaining the changes.

Elizabeth began to envision herself as being more flexible in project planning and problem solving. Then she held a series of small group meetings with members of her staff to brainstorm ways they could work together to set common goals and solve scheduling problems. Out of these meetings grew a plan for a twenty-minute weekly meeting to discuss the goals and schedule for the week. Facilitation of this meeting would rotate among team members, so that everyone would have the chance to lead the process. After six months, she asked for feedback from her staff about how the new plan was working. Staff members expressed enthusiasm for the change in how the department was being managed. "I like feeling like a member of a team working together to achieve a common goal," one developer told her.

Over a long weekend, Elizabeth took the time to evaluate her own response to the change. The email she sent me at the conclusion of this self-monitoring process gratified me greatly: "Thanks for helping me to move toward becoming the manager I always wanted to be," she wrote. "I used to feel as if I were driving a team of horses, pushing and pulling to make them work together to meet project deadlines. Now it feels as if the responsibility for getting things done is shared by everyone on the team. Some of the methods team members have come up with to solve scheduling problems are things I would never have thought of, but they have worked beautifully. I guess I'm learning to be more flexible and discovering that being a manager doesn't mean running the whole show."

At our next coaching session, I worked with Elizabeth on devising a plan for maintaining her new style of flexible management. I asked Elizabeth how she had felt when she was managing her group as if she were driving the team of horses. She described the feeling as one of overwhelming strain. Then I asked how she felt when she was flexibly collaborating with her team. She described that feeling as one of ease and relief. I asked Elizabeth to picture a series of other situations in which she would characteristically feel overwhelming strain, including situations away from the office, and then another series in which she would feel ease and relief. By visualizing the experiences of strain and relief in various situations, Elizabeth became so familiar with the feelings that she could step into them at will in a few seconds. This familiarity is the key to maintenance, as she would need the ability to discriminate immediately when feelings of strain began to emerge. The feeling would signal Elizabeth that she was reverting to inflexible behaviors.

"If you begin to experience strain," I told her, "shift back to step four in the process and devise a new set of action steps to return to the feeling of relief." Each time she acted on the internal cues to reinstate her desired behavior, the feelings of ease would return. Experienced a few times, this pattern of self-monitoring and constructive action in response to internal feedback constitutes an immense reward. It helps us strengthen and sustain new behavior patterns.

After completing her second year on the job, Elizabeth once again participated in the 360-degree analysis. Her inner sense of accomplishment was validated by results that documented approval of her new flexible and collaborative management style. Through attentiveness and commitment, Elizabeth had made a significant change that benefited her personally, as well as her company.

Adopting the Wisdom Perspective

Responding constructively to feedback by making changes in our behavior is easier when we adopt what is called the wisdom perspective. Researchers conducted a series of four studies involving 533 people who ranged in age from 20 to 89 years, and who represented diverse educational and socio-economic backgrounds. While the overall results did not correlate wisdom with age, the top 20 percent of those ranking high on wisdom-related tasks included more older participants than younger ones.[14]

Participants in the study were asked to respond to a series of situational questions. For example: "A fifteen-year-old girl wants to get married right away. What should one/she consider and do?" The following response was scored as low on the wisdom-related scale: "A fifteen-year-old girl wants to get married? No, no way, marrying at age fifteen would be utterly wrong. One has to tell the girl that marriage is not possible." After further probing, the respondent continued, "It would be irresponsible to support such an idea. No, this is just a crazy idea."

The following response was scored as high on the wisdom-related scale: "Well, on the surface, is seems like an easy problem. On average, marriage for fifteen-year-old girls is not a good thing. But there are situations where the average case does not fit. Perhaps in this instance, special life circumstances are involved, such that the girl has a terminal illness. Or the girl has just lost her parents. And also, this girl may live in another culture or historical period. Perhaps she was raised with a value system different from ours. In addition, one has to think about adequate ways of talking with the girl and consider her emotional state."

From this response, we can begin to deduce some of the qualities that make up the wisdom perspective. The second speaker displays psychological mindedness that leads to high empathy for the girl's situation. Moreover, the response indicates flexibility, a willingness to remain open to new information, and a nonjudgmental emotional attitude. These same qualities can help us make needed changes in our lives in response to feedback. Wisdom seems to require that we balance two aspects of ourselves: our observable or local self, apparent as we go about our daily lives, and our essential self, who is wise, vast, and already complete. Many of us tend to view the local self as the real self, and either disregard the essential self, or pay attention to it only on spiritual occasions. When we have made a mistake, especially one that causes us to receive negative feedback, we can be so consumed with making changes to the local self, that we forget that we also have an essential self who already possesses all the wisdom we need to know about how to rectify the situation.

In my work with clients, I try to help them maintain the wisdom perspective. I hold a mirror for them to view the resources of their essential self and the vast energies available to them as a result. Doing so helps them to understand that their stumbling and the bumbling behaviors are not all that they are. In time, they learn to laugh at themselves. Critical self-judgment has no place in the wisdom perspective. Rather, problems are transformed into opportunities for bringing forth the essential self to a greater degree. Making positive change requires a balanced perspective on the self. I encourage my clients to view their disconcerting behavior in its fullness, without self-deception, while simultaneously being aware of the wise essential self, always accessible. This balanced viewpoint allows them both to acknowledge their error and to claim their greatness, utilizing the wisdom of the essential self to initiate and sustain change.

As I also remind my clients, it's important to hold this balanced perspective in mind when they are giving feedback to others. The local self is very concerned with how it looks and can thus be very defensive. Harsh or overly critical feedback brings the local self to the fore, ready to do battle. Feedback that concentrates on the behavior, clearly differentiating it from the person who has committed the error, is much more valuable. Such feedback helps the person receiving it to remember that while the local self can make mistakes, the essential self is wise enough to admit those mistakes and make positive changes.

Coaches aren't therapists. I do not counsel people with severe behavioral problems. Rather, I see people who are dissatisfied with some aspects of their lives and want to make changes. In our work together, I encourage my clients to revisit those times in their lives when they experienced a deep sense of fulfillment, for at those times they were closer to their essential selves. Establishing a strong connection with the essential self is the foundation for transformative change. My background in academia has drawn several clients to me. Many are eager to move out of academia, and they come to me because they know I have been through this process. As was true for me, these people often find it difficult to envision themselves in another role. Through exercises, such as ones in this book and sessions of one-on-one feedback, I help these clients bring to consciousness the skills and activities that most fulfill them, and to envision ways to use these skills as the basis for a new life.

Anita, a part-time adjunct instructor at an art institute, came to me because she was dissatisfied with her career in academia. Although Anita held a Ph.D. in art history and architecture, she felt taken advantage of, marginalized, and underpaid. At our first meeting, I asked Anita to imagine a circumstance in which she would feel as though she were an integral contributor, compensated equitably, and valued for her abilities. As we talked through various possibilities, Anita came to realize that she found the structured environment of academia restrictive and not nourishing. What she really wanted was a way of expressing her skills and talents that was more free-flowing and energetic.

The next step was for Anita to announce to friends, associates, and colleagues that she was looking for a new position. In a very short time, she was offered a job as the head of a foundation devoted to promoting the work of a foreign artist, who was, by that time, deceased. The artist's paintings were largely architectural in content, and Anita's expertise was especially relevant to understanding and explaining the artist's work to others. She now represents the artist to galleries, and explains his focus to the world through catalogs, publications, and gallery showings. The foundation she heads also contributes funds to help the disabled in her community. Key to Anita's new life was holding the vision of finding work that would allow her to use her talents fully. The wisdom perspective allowed her to envision ways to actualize herself and contribute to her community.

Like my client, you have the perspective of your essential wisdom self available to you. The following exercise can help you use it to make changes that have a positive impact on you and your community.

FEEDBACK FROM THE WISDOM PERSPECTIVE

Choose a place for this exercise that promotes a feeling of connection with your wise essential self. For instance, you may want to do this exercise outdoors, where you can view the vast sky to help promote this connection. Take along a book of poetry, other writings, paintings, or music tapes—whatever helps you connect with the wisest part of yourself.

Make yourself comfortable. Take a few very deep breaths, filling your lungs with air. Slowly, release the air, and with it, any tensions, physical or mental. Simply allow them to flow away with your out-breath.

Read the poems you brought, view the paintings, or listen to the music as a way of connecting with the most wise and nurturing part of your self. Once you feel connected to the deep being within, stay there a while. Allow the comfort and nourishment of this part of you to permeate all of you.

[] Now recall a circumstance in which you received feedback, either from an inner knowing about some aspect of your life that is not working, or from an outside source.

[] Consider the feedback and ask yourself, Would making the changes suggested bring me closer to expressing my essential self?

[] Maintaining your connection with your essential self, engage it in conversation. Ask that vast, resourceful self within to help you envision ways you could more fully express all that you are. Take time with this process. What circumstances would promote your greater self-expression? In your imagination, invite your essential self to tell you what you need to do.

[] Listen, listen for the messages, without censoring. Write down whatever thoughts or ideas come.

[] Before leaving this place, determine what steps you will take in response to the messages that you have heard.

[] The process is not complete until you take action. Do something that opens the door for your essential self to come more fully into your life.

[] Continue to listen. Each time you take a next step, listen for another directive. Keep the connection with your essential self vital by continuing to ask for inner guidance as you make changes.

[] Pay attention to how you feel as you respond to the inner advice you are getting. You'll know you're on the right track if each action you take gives you an increasing sense of ease and flow.

This process can be used to help you give skillful feedback to others, as well, such as key members of your community, family, or your staff. Simply change the question, Would making the changes suggested bring me closer to expressing my essential self? to Would making the changes suggested bring the key members of this group closer to expressing the group's core values? As the various communities of which you are a member come to express their core values more fully, you also bring forth that part of you that contributes to your own well-being and that of society.

CHAPTER NOTES

1. Robert V. Farese, "Insulin-Sensitive Phospholipid Signaling Systems and Glucose Transport. Update II1," *Proceedings of the Society for Experimental Biology & Medicine* 226, no. 4 (2001): 283–295.

2. Mary H. McCaulley, "Myers-Briggs Type Indicator: A Bridge Between Counseling and Consulting," *Consulting Psychology Journal: Practice & Research* 52, no. 2 (1996): 122–132.

3. Manuel London and James W. Smither, "Feedback Orientation, Feedback Culture, and the Longitudinal Performance Management Process," *Human Resource Management Review* 12, no. 1 (2002): 81–100.

4. Ira Chaleff, *The Courageous Follower* (San Francisco: Berrett-Koehler Publishers, 1998).

5. Gordon W. Blood, "A Behavioral-Cognitive Therapy Program for Adults Who Stutter: Computers and Counseling," *Journal of Communication Disorders* 28, no. 2 (1995): 165–180.

6. Kimberlee J. Trudeau and Rosandra Reich, "Correlates of Psychological Mindedness," *Personality and Individual Differences* 19, no. 5 (1995): 699–704.

7. See note 6 above.

8. University of Washington, "Neuroscientists Searching for Roots of Empathy Find Brain Regions Involved in Learning by Imitation," news release, January 22, 2002, http://www.washington.edu/newsroom/news/2002archive/olozarchive/kol2202.html (accessed September 22, 2003).

9. Joseph V. Ciarrochi, Amy Y. C. Chan, and Peter Caputi, "A Critical Evaluation of the Emotional Intelligence Construct," *Personality and Individual Differences* 28, no. 3 (2000): 539–561.

10. Joan F. Brett and Leanne E. Atwater, "360 Degree Feedback: Accuracy, Reactions, and Perceptions of Usefulness," *Journal of Applied Psychology* 86, no. 5 (2001): 930–942.

11. See note 10 above.

12. Paul B. Baltes and Ursula M. Staudinger, "Wisdom: A Metaheuristic (Pragmatic) to Orchestrate Mind and Virtue Toward Excellence," *American Psychologist* 55, no. 1 (2000): 122–136.

13. Leonard A. Jason et al., "The Measurement of Wisdom: A Preliminary Effort," *Journal of Community Psychology* 29, no. 5 (2001): 585-598.

14. See note 13 above.

CHAPTER II

provide and accept support

SOME PEOPLE BECOME GREAT because they believe they can; others, because someone else believes they can. When we cannot believe in ourselves, we may be able to believe in someone else's belief. With the support of others, we can often reach for accomplishments that seem beyond our individual capability.

Support abounds within the body. Consider the skeletal system. Without the support of muscles, our bones would be piled in a stack, leaving us incapable of standing upright or moving. Tendons—rope-like connective tissues—attach muscle to bone, especially important across joints that might tear muscles during movement. The simple act of standing requires that muscles, attached to bones in just the right places, are stimulated by neurotransmitters released from nerve endings to contract periodically so as to support our upright position and resist gravity.

Walking requires an even greater degree of coordination between nerve, muscle, tendon, and bone. The cascade of synchronized events required for walking begins with a thought in the motor cortex of the brain. Neural impulses originating from the upper motor neuron in the cortex travel downward through the spinal cord to synapse with lower motor neurons in the lumbar and sacral regions of the spinal cord. Axons from these spinal neurons bundle together and leave the spinal cord as nerves that contact muscles in the foot, legs, thigh, and hip. The nerve endings synapse with these muscles and release neurotransmitters that cause the muscle fibers to contract. A single nerve contacts several different muscles and coordinates their contractions. As we walk, upper and lower motor neurons stimulate, in a coordinated fashion, each of 118 flexor and extensor muscles—58 in the right foot, leg, thigh, and hip, and 58 in the left. These muscle contractions allow our bones to move in a coordinated way. Walking is thus made possible because of the structural support and dynamic interrelationship between many physiological systems. Acting alone, neither bones, muscles, tendons, nor nerves could fulfill the purpose of allowing us to move. Only interdependent support between various systems makes movement possible.

Providing Effective Support, One-on-One

When this interdependent support among elements of the body breaks down, severe dysfunction can result. For instance, damage to the lower motor neurons in the spinal cord from accident or illness can compromise our ability to stand and walk. Many of us remember being frightened about polio as children. This virus finds its way into the lymph nodes and blood, and eventually reaches the lower motor neurons of the spinal cord. The virus damages the neurons in the lumbar and sacral regions so that they no longer stimulate the muscles of the hip, thigh, leg, and foot. As a result, walking—even standing without support—becomes impossible.

Polio reached its peak in 1952, when nearly 58,000 cases were reported around the world, nearly a third of them in the United States. That year, American medical pioneer Dr. Jonas Salk inoculated himself, his wife, and their three sons with a vaccine containing an inactivated polio virus. No one became ill. The next year the vaccine was tested nationwide, and Salk became famous. A few years later, another American researcher, Dr. Albert Sabin, developed a vaccine that contained a weak strain of the live polio virus. Sabin's vaccine could be taken orally, an advantage in third world countries in which the disease was raging. Tested worldwide in 1957, the oral vaccine, which conferred lifelong immunity, soon replaced Salk's vaccine. Sabin's oral vaccine virtually eliminated polio in the western hemisphere. In 1988, the World Health Organization adopted a resolution to eradicate polio throughout the world by the year 2000. Just nine years later, the number of cases reported worldwide had decreased by 90 percent.

Many public figures contracted polio prior to the development of the vaccine, including President Franklin Delano Roosevelt. Today, violinist Itzhak Perlman reminds us of the debilitating effects of this disease, as he sits with crutches besides him on stage. Innovative educator Frederick Hudson, founder of the Fielding Institute, was also among those who contracted polio before the vaccine was developed. He relates the story of the imaginative and creative support that a nurse provided him as he lay in his hospital bed in 1943, nine years old and unable to move. He remembers his nurse, Susan, telling him, "Your future, Frederick, is hidden on the ceiling, and you can find it if you look very hard. Look for what you will be doing as you grow up. It's all up there . . . Frederick; all you have to do is study the ceiling. When you see your future, it will start to happen!"[1]

With little choice but to do as Susan suggested, Hudson looked at the ceiling and saw himself running and playing, "bouncing through the woods, like a gazelle, alive with movement." After a while, he felt a tingle in his toes. Nurse Susan told him, "You are now in training, so practice moving your legs . . . soon you will walk and then you will run."

Susan attached strings to Hudson's toes so that he could ring a bell if he wanted her. Soon he was making so much noise that the other nurses complained. Susan coached Hudson to visualize himself walking and encouraged him until he was able to do so. She turned his hospital room into a rehabilitation gymnasium. Before long, the other nurses joined the program, remarking on his progress and congratulating Hudson on his efforts. Several months later, Hudson *walked* out of his hospital room. Nurse Susan's support encouraged him to use limbs paralyzed by polio until he believed he would walk and ultimately regained his ability to do so. Although he may not have believed it was possible, Hudson believed in Susan's belief that he would walk again.

Regaining the ability to walk required Hudson's body to grow new nerves that connected undamaged lower motor neurons in his spinal cord to the muscles of his legs. Hudson's story raises an interesting medical question: Could his belief that it was possible for him to walk again facilitate this growth? Though we cannot answer this question definitively, there is evidence that psychological factors, such as belief, can affect a patient's physical healing. Physicians have long known that patients who believe a drug they are taking will help cure them have better medical results, even if the drug they are taking is only an unmedicated placebo, such as a sugar pill. In fact, some clinicians have urged physicians to think of the placebo effect as an instance of "remembered wellness," and to harness its power to help patients heal.[2]

One article, published in a national medical journal, points to three criteria necessary for patients to remember wellness: First, patients must have positive beliefs and expectations; second, their physicians or health care professionals must also have positive beliefs and expectations; and third, there must be a good relationship between both parties.[3] Could the placebo effect have helped Hudson's healthy motor neuron axons grow by sending new axons to contact his lower limb muscles, thereby allowing him to wiggle his toes? Applying the remember wellness criteria to Hudson's case, we see that the young boy's vision of himself running and playing in the woods is evidence of his own positive beliefs and expectations. Nurse Susan tying the bell to Hudson's toe demonstrates that she shared these beliefs and expectations. And, the fact that Hudson recounts this incident in a book written years later testifies to the good relationship between the two.

Hudson's optimism may also have aided his recovery. Current research indicates that optimists enjoy longer life, while pessimists display high risk factors for early death. One recent study, in particular, has suggested three possible reasons for correlating optimism with longer life: (1) optimists are less likely to develop depression and learned helplessness; (2) they are likely to have a more positive attitude in seeking and receiving medical help, and less likely to blame themselves or assume the worst outcome; and (3) their positive attitude might actually lead to positive changes in their immune system.[4] While the causes of Hudson's complete recovery can never be known for certain, the positive outcomes of nurse Susan's support are clearly demonstrable. Hudson returned to school and eventually earned a doctorate degree from Columbia University. Recognized as an expert in adult change, Hudson founded the Fielding Institute, an innovative doctoral studies graduate program. Later, he established The Hudson Institute of Santa Barbara as a training center for professionals focusing on "renewal and resilience" at work and at home. Currently, he teaches others to mentor and coach, having experienced for himself, early in life, the inestimable value of support.

Qualities of a Personal Mentor

What qualities make someone able to provide helpful support for others? A two-part study, related to the work on the wisdom perspective that we looked at in chapter 10, measured wisdom as the quality that allows people to be effective mentors for others.[5]

In the first part of the study, a group of forty-three people was asked to name the wisest living person they knew. Some respondents named former teachers, friends, or family members. Others named world leaders, such as former President Jimmy Carter, the Dalai Lama, or Mother Teresa; spiritual authors, such as Elizabeth Kubler-Ross and Shakti Gawain; political leaders, such as Nelson Mandella and Barbara Jordan; business leaders, such as Steven Jobs; a physicist, Stephen Hawkins; and an entertainer, George Burns. The respondents were also asked to give an illustration of an incident that revealed this person to be wise, to delineate qualities that made this person wise, and to describe the effect that this person's wisdom had on their lives. The researchers then took qualities culled from the responses to these open-ended questions and used them to develop a scale designed as a measurement tool for wisdom, called the Foundational Value Scale (FVS).

In the second part of the study, the researchers tested the FVS using 140 college students at DePaul University. As in the first part of the study, the students were asked to name the wisest living person they knew. Their answers reflected the same categories and types as those chosen by participants in the first part of the study. Interestingly, 69 percent of the students had met their nominated wise person in an educational, family, or work setting. Then the students were asked to evaluate the qualities of the wise person they had named using the FVS. For each quality listed on the scale, the students circled a number from 5 to 1, in which 5 indicated that the person "definitely" had a particular quality and 1 indicated that the person did "not at all" have this quality.

Researchers analyzed the results of the two studies and identified five factors associated with wisdom: harmony, warmth, intelligence, reverence for nature, and spirituality. *Harmony* included the qualities of balance, positive self-esteem and self-love, good judgment, the capacity to cope with uncertainty, the ability to see meaning and purpose in life, a sense of gratitude and appreciation, the ability to experience an underlying unity in life, an appreciation for things as they are without embellishment, and an openness that can accommodate whatever experiences arise. *Warmth* encompassed humor; kindness and compassion; animation, as reflected in rapture, joy, hope, and happiness; and the quality of living in the present. *Intelligence* related not only to genius, but to how it was used to solve real-life problems and to help others. *Reverence for nature* included concern for the health of the environment, a sense of childlike wonder and awe, and the capacity for flow—being so involved in an activity that nothing else seems to matter. *Spirituality* involved feeling love, fellowship, or union with God and living a spiritual life.

Another researcher identified three energetic patterns that he considers to be basic within the universe: differentiation, subjectivity, and communion. *Differentiation* encompasses everything that is outside ourselves—all the abundant expressions of life and creativity in the world.[6] *Subjectivity* refers to the universe within—a person's unique interior depth and spontaneity. *Communion* reflects the relationship between the two, characterized by a sense of interrelatedness and community. Of the five factors revealed by the FVS as aspects of wisdom, differentiation is reflected in the association of wisdom with reverence for nature, including spontaneity and creativity; subjectivity finds expression in the various qualities associated with inner harmony; while communion is expressed in the qualities of warmth toward others and helpful intelligence (using one's intelligence to help others).

A better knowledge of wisdom, the researchers concluded, could help to provide the potential for introducing vital, energizing symbols and messages that could be used to restore a sense of meaning to those who are discouraged and alienated. Further, wisdom could help to develop and sustain community leaders, who need to display resiliency in the face of oppression, a broad ability for coping with uncertainty, and the aptitude to frame events in a larger context.

Supporting Others with Wisdom

The qualities of wisdom are relevant to the many opportunities we have to support others in everyday life, in our family, workplace, and community. I think of a family story told to me by one of my coaching clients. Sally's daughter Debbie became pregnant while she was in high school. Though she was personally distressed and disappointed, Sally resolved to put her own feelings aside so that she could act as a wise mentor to Debbie and her daughter's boyfriend, Mac. Sally encouraged the young couple to explore their feelings, separately and together. As a result of these discussions, Debbie and Mac decided that although they wanted to continue their relationship, they were not ready to marry. Nor did they want Debbie to have an abortion or to give the baby up for adoption. Moreover, Mac wanted to participate in raising his child.

With these decisions as the foundation, Sally called a family meeting. In addition to her own parents, Sally invited Mac's parents and grandparents. When all were assembled, Sally opened the meeting by describing the situation as one that involved not only Debbie and Mac, but everyone in both families. Everyone present agreed to honor the decisions Debbie and Mac had made about how to proceed. The families discussed how they might share responsibility for the coming child while Debbie and Mac completed their schooling. Shared responsibilities were identified, and the difficulties family members might meet in following the plan were explored and discussed. As Debbie's time to deliver the baby approached, Mac remained beside her, supported by members of both families.

Most would agree that Sally is an extraordinarily wise woman. Rather than blaming Debbie or Mac, or bemoaning the unfortunate circumstance, Sally remained focused in the present and applied her intelligence to solving the problem at hand. Her decision to honor the choices Debbie and Mac made demonstrated warmth, compassion, and a sense of harmonious balance. She did not propose taking over responsibility for raising Debbie and Mac's child, but demonstrated good judgment by keeping her own needs, as well as the needs of the child, in mind. Her solution reflected an appreciation for the interconnectedness evidenced in nature; she realized that the problem did not belong to Debbie and Mac alone, but involved members of both families. The two families' support for each other built a strong foundation of support for the new family of Debbie, Mac, and their child.

Take a moment to ask yourself how you might have reacted if Debbie were your daughter or Mac, your son. Ask yourself what opportunities you have had to use the qualities of wisdom in solving a problem with which you have been confronted? How did you do?

Organizations that Provide Support

Like individuals, organizations benefit from the ability to provide effective support. Psychologists say that a "hardy" organization is characterized by its ability to transform a stressful circumstance into an opportunity for positive change by taking decisive action to solve the problem.[7] Pivotal to this ability is an organization's willingness to give assistance and encouragement to individuals and groups within the organization to help them manage

stressful circumstances. This support consists of actions that value cooperation, credibility, and creativity.

Holding the qualities of a hardy organization in mind, let's look at an
example of a corporation known for its innovative solutions to technical
problems. The computer chip manufacturer Intel demonstrates well how a
commitment to organizational support can transform stressful circumstances into opportunities for positive change.[8] Despite a slump in the computer industry, Intel has set out to develop the world's smallest and fastest
computer chip. The technical challenges that need to be overcome to meet
this goal are considerable.

The development process began with setting a clear goal. "In a few
years," Youssef El-Mansy, Intel's head of logic-technology development told
a small team of scientists, "I'd like you to publish a paper that says, 'We have
designed the fastest transistor in the world.'"[9] Several things are significant
about El-Mansy's statement. First, he issued his challenge to a carefully
chosen team of one hundred researchers rather than to individual scientists,
thus encouraging a culture of cooperation. Moreover, the research team consisted of new Ph.D.'s who had just been hired by the company. As El-Mansy
explained, "They have this great advantage of not knowing what's impossible or what's too hard to be worth trying. So they try it anyway. And sometimes it works."

El-Mansy's assumption that the research team would take on the full
responsibility for the chip's development created a climate of credibility.
El-Mansy also did not suggest how the team might solve the technical problems involved in building the fastest chip. Rather, he created conditions in
which creativity could flourish. For instance, in an attempt to knock down
barriers between the research and development team and manufacturing,
Intel housed the researchers adjacent to a chip manufacturing facility,
where new ideas could be tested quickly to provide a reality check. Moreover, Intel's top executives did not micromanage the project. Whenever the
team met with serious bottlenecks, the managers provided extra people
with the right skills.

Hardy organizations, according to researchers, are characterized by a
belief in change and its value, which keeps them anticipating the future and
developing the products and services to turn it to advantage.[10] Based on the
evidence, Intel seems to fit this mold. Key to its success is the ability and
desire of Intel's management to provide the right kinds of support to sustain
innovation.

WHAT SUPPORT CAN YOU PROVIDE?

Recall the exercise you did in chapter 9, in which you explored your social network as a foundation for social action. In this exercise, you will have the opportunity to consider the kinds of support you can and do provide for individuals and groups within your social network.

[] Bring to mind several people and/or groups in organizations within your social network to whom you have provided support within the last six months.

[] Review the circumstances of that support. Were the individuals and/or groups in a period of transition? A period of expansion? A time of fear or loss? A time of new responsibilities or acquisitions?

[] What kinds of support did you provide for the people and/or groups? What specifically did you say or do?

[] How would you characterize the support you provided? Recall that the Foundational Value Scale lists five qualities of wise support: harmony, warmth, intelligence, reverence for nature, and spirituality. In what ways did the support you provided reflect these qualities?

[] If the support you provided was lacking in one or more of these qualities, envision a different scenario. Visualize yourself providing support characterized by kindness and compassion, as you intelligently help others solve their problems. Envision the scene unfolding before you and imagine what happens.

[] Now ask yourself: What circumstances exist within my social network that might benefit from additional support? Who might benefit from such support? What kinds of support would be most helpful?

[] Consider several possibilities for future action and choose one that you feel drawn to. Allow your imagination to explore the scenario fully.

[] What's your next step? Determine when and where you'll take it.

[] If no possibility emerges, set your intention to be alert and to take action when a circumstance arises in which you might provide beneficial support.

Asking for and Accepting Social Support

As we saw at the beginning of this chapter, walking entails a coordinated system of support between nerve, muscle, tendon, and bone. As you recall, neural impulses, originating in the brain, travel down the spinal cord and out to the muscles of the foot, legs, thigh, and hip. There, nerve endings synapse with the muscles and release neurotransmitters that cause the muscle fibers to contract, making walking and other movements possible. Recent research has focused on understanding the development of the relationship between neurotransmitters and the skeletal muscles they stimulate.

Three researchers have written about the development of the relationship between nerve and muscle, and the role of the neurotransmitter acetylcholine, which is released from nerve endings at the neuromuscular junction.[11] A muscle is able to respond to stimulation by acetylcholine because of chemical receptors within the muscle specifically matched for this neurotransmitter. These receptors are generated during the process of fetal development, before the nerve ever contacts the muscle. At this stage, the receptors cluster in the middle of each muscle fiber. As development proceeds and the nerve reaches the muscle, the patterning of the acetylcholine receptors is refined to maximize the amount of simulation of the muscle fibers.

From these observations, the authors conclude that "interactions between nerve and muscle proceed in two directions in that each cell prepares for this exchange before contact." The possibilities that follow from this conclusion are intriguing. Might it also be that the muscle generates acetylcholine receptors and positions them as an *invitation* to the nerve to stimulate it?

Hope as an Invitation to Social Support

As between muscle and nerve, human support interactions proceed in two directions. As we saw in the case of Frederick Hudson, his belief in his nurse Susan's belief that he would walk again seems to have contributed to his recovery. Might the emotion *hope* function in a similar way?

Two nurses studied the concept of hope by focusing on several populations of patients in need of support, including those awaiting heart transplants, those with spinal cord injuries, and those with breast cancer.[12] They found that hope is often the only thing that sustains heart transplant patients day-to-day. Such patients are often desperate, knowing that a transplant is their only chance to lead a normal life. Often, they suffer extraordinary setbacks. One patient who had been anesthetized in anticipation of his second heart transplant woke up to discover that the operation had not been done! Evidence shows that heart transplant patients keep hope alive despite such setbacks by soliciting social and emotional support from hospital staff, family, clergy, and friends. They also reach out to former transplant patients, with whom they exchange positive survival stories. As is true with muscle receptors and the neurotransmitters that stimulate them, the interactions between heart transplant patients and the people who support them flow both ways. The patients' hope causes them to reach out to others; because of this hope, psychological and emotional support flows to the patients.

A similar dynamic exists for patients with spinal cord injuries. Realizing their limitations, these patients evaluate their progress by aiming for small gains. They maintain hope by noting the slightest increase in mobility. One patient commented, "Every little step I took was more important to me than what I had in the end."[13] Like heart transplant patients, spinal cord injury patients reach out to physical therapists, nurses, and staff to support them in this process. Again, we might conclude that they reach out to receive support because of hope, and their hope for small gains invites support from the people around them.

Support groups such as Alcoholics Anonymous, groups for people with cancer and other physical and emotional conditions, and groups for people in transition, such as those recently widowed or divorced, are an important aspect of social support for many people. Breast cancer patients understand that pain, suffering, and death are real possibilities in their lives. They carefully attend to any bodily changes, examining themselves for new lumps and monitoring signs of aging, which they often perceive as being accelerated by the treatments for their cancer. They live with a sequence of hopes. If a needle biopsy is needed, they hope it shows that their breast lump is a fluid-filled cyst and not a malignant tumor. If surgery is needed, they hope it will entail a lumpectomy and not a mastectomy. At each checkup, they hope that nothing suspicious will be detected. Their vulnerability makes these patients especially good at assessing available resources, including carefully evaluating people as potential sources of support. Most patients actively seek connections with other women who have had breast cancer, including joining breast cancer support groups. Support from others with breast cancer and selected family members helps these women in their fight to control negative and fearful thoughts.

As a result of these studies, the two nurses arrived at the following definition of *hope:* "Hope is a response to a threat that results in the setting of a desired goal; the awareness of the cost of not achieving the goal; the planning to make the goal a reality; the assessment, selection, and use of all internal and external resources and supports that will assist in achieving the goal; and the re-evaluation and revision of the plan while enduring, working, and striving to reach the desired goal." As this definition makes clear, one of the necessary conditions for the positive outcomes hope makes possible is the ability to ask for and receive social support.

Two other researchers came to a slightly different conclusion about the importance of hope when they asserted that hope requires an individual to believe that improvement is possible.[14] However, this study concluded that while both hope and social support were associated with reduced distress, hope was not the mechanism through which social support and its positive outcomes was invited. The participants in this study were 111 mothers of children, ages five to eighteen, who had cerebral palsy, spina bifida, or insulin-dependent diabetes. The mothers' degree of hope was evaluated by an inventory called the Hope Scale; the level of social support they received was evaluated by the Social Support Questionnaire; and the degree of dysfunction of their children was evaluated by the Brief Symptom Inventory.

In evaluating the results of these measurement tools, the researchers concluded that mothers caring for children with severe disabilities did not experience more distress than mothers caring for children with lesser disabilities. Rather, distress was reduced in mothers who scored high on the Hope Scale and who received high levels of social support. Social support— the perception of the number of people available and the degree of satisfaction the mothers experienced as a result of the support they received—was correlated closely to reduced distress.

However, quite apart from social support, hope itself was associated with reduced distress. The Hope Scale revealed that people with higher levels of hope have more goals. Faced with obstacles, they continue to pursue these goals by generating other strategies. Moreover, people with higher levels of hope have a sense of optimism about outcomes, a perception of self-control, greater problem-solving ability, enhanced self-esteem, and the ability to use humor to cope with stressful events. The authors concluded that both factors, hope and social support, independently contribute to reduced distress. It may be, however, that hope is the preexisting condition that allows people to open up to receiving support from others.

Remember Sally and her pregnant teenage daughter, Debbie? Sally's wise, one-on-one support, coupled with access to a network of social support from various groups, contributed to a positive outcome for Debbie, Mac, and their child. Debbie's high school counselor put her in touch with a support group for single teenaged mothers-to-be. The first meeting was scary, but soon Debbie was soon talking freely with the other pregnant teenagers, sharing hopes and fears and exchanging useful information. This group was important to Debbie in helping her handle the stress of transition from being a pregnant high school student to being a mother attending college part-time. A tip from a member of the support group led Debbie to discover that a nearby college, Boston University, had a special program for student parents. She also learned of a group for unmarried fathers. Mac attended the group and found that many young men shared his values and wanted to help raise their children. The support of this group helped Mac become more comfortable with his situation and with the prospect of being a single father. Debbie and Mac also began attending birthing and new-parent classes. Throughout this process, the young couple's social network grew, and they found a wide range of support resources.

Epidemiologists and psychologists have documented evidence that social support given to pregnant women has a positive effect on their newborn infants.[15] They studied 225 women who received care in the prenatal clinic of a university-affiliated hospital and a university-affiliated, low-risk birth center. These women also got support from family, the baby's father, and other resources. The researchers suggest that this support contributed to lower levels of the stress hormones that can cause low birth weights or premature births. They found that social support may also lead to healthier lifestyles for pregnant women, and increase the likelihood that they will seek health-related information.

Debbie and Mac's son was born at full term and weighed a healthy seven and one-half pounds. Debbie and Mac were both able to attend Boston University part-time, while holding down jobs and sharing in the care for their son. Though the couple decided not to marry, both continue to be involved in their son's life, as are their parents and other family members.

The Benefits of Providing and Accepting Support

Who benefits from support? In fact, those who give support may benefit even more than those who receive it. In one study, recruited nonprofessionals with multiple sclerosis were trained in active listening and taught to provide support to sixty-seven other people with multiple sclerosis over a two-year period.[16] The peer supporters were paid ten dollars an hour to provide telephone support for fifteen minutes a month throughout the two-year period of the study.

The peer supporters were recruited through several means. Some had volunteered to participate in a pilot group and were selected because they demonstrated an ability to communicate with others in the group and were willing to commit to a two-year effort. The Massachusetts chapter of the National Multiple Sclerosis Society recommended others because they were helpful in volunteer activities. Peer supporters completed a questionnaire before beginning to give support, and again after one and two years of providing support. The sixty-seven participants who received support also completed questionnaires at the same intervals.

Three years after they finished providing active support, the peer telephone supporters were queried about the changes they noticed over the course of their participation in the study. In this follow-up evaluation, the peer supporters reported improvement in more areas as compared to improvements experienced by the patients who received the support. For instance, the peer supporters were seven times more likely to feel a sense of well-being than were the patients receiving support. They reported improved listening skills, a stronger awareness of the existence of a higher power, increased self-acceptance, and enhanced self-confidence. Peer supporters also reported experiencing a sense of inner peace that allowed them to listen to others without judgment or interference.

Interestingly, these aspects of well-being accelerated during the second year of the study. As the supporters became more effective and more outer-directed, a shift occurred in the way they thought about themselves. Participants who received support exhibited change in a number of these areas as well, but the changes were less pronounced than those experienced by the people who provided the support.

What brought about these positive changes for people giving support to others? The authors of the study propose that these shifts occurred because of the number of personal stories the peer supporters exchanged with the people they supported. As they heard more and more stories of other people with multiple sclerosis facing challenges, the supporters were able to disengage from their usual ways of thinking about themselves and their condition. In other words, as their focus moved from concern about themselves to concern for others, their attitudes about their own multiple sclerosis altered as well. As one supporter commented: "It's tough to get depressed because you're helping someone." Another supporter explained the change this way: "There's a quietness when I'm talking to someone, and I'm really listening to them. I have to make an effort not to try and top them. It's gotten easier. And I can listen, and I become interested in what he's talking about. That's a change. There's a quietness in the soul because of it."

Of course, people who receive social support also benefit in many important ways. Another recent study suggests that our continued ability to think, remember, and create as we age is influenced by the amount of emotional support we receive. Researchers studied 1,313 high-functioning men and women, ages seventy to seventy-nine, over a period of more than seven years to explore the relationship between cognitive ability—the ability to reason, use abstract language, and remember details—and the amount of emotional support elderly people receive from family and friends.[17] Participants in the study were asked several questions: How often do your spouse, children, close friends, and relatives make you feel loved and cared for? How often do they help with daily tasks like shopping, giving you a ride, or helping you with household tasks? How often do they give you advice or information about medical, financial, or family problems? Then the cognitive functioning of these participants was assessed using a battery of tools to reflect detailed and subtle changes.

The researchers found that greater emotional support was associated with better cognitive functioning, regardless of the physical health, physical activity, or socioeconomic status of the people receiving the support. In fact, the more often participants received emotional support, the more slowly their cognitive functions declined over the seven years of the study.

The benefits of social support can be summarized in the larger context of cumulative research over the last twenty-five years.[18] Communities characterized by a cooperative network of social relationships are linked to greater health and longer life for community members. People trust and help one another. Their communities distribute information promoting healthy behaviors, and people see the value in adapting such behaviors to their own lifestyles. Members of these communities also have easier access to health-related services than do members of other communities.

BENEFITS OF GIVING AND RECEIVING SOCIAL SUPPORT

Settle in a comfortable place where you will not be disturbed and reflect on the following situations:

[] Recall a situation in which you asked for and received support. What were the circumstances?

[] What support did you ask for and what support did you receive? What benefits did you derive from asking for support? From receiving support?

[] What difference did asking for and receiving support make in your life?

[] Recall a situation in which you gave support. What were the circumstances? What support did you give?

[] Did you derive benefit from providing support? If so, what was it?

[] When did you become aware of the benefit you received by providing support?

[] How did giving this support impact your life?

[] Now bring your awareness to the current time. Think about your family, friends, and associates—the people in your social network.

[] Allow your attention to drift from person to person. As you bring each person to mind, ask yourself, Is this someone who could benefit from my support at this time?

[] When you have completed this process, ask yourself: Is there someone who stands out as someone I feel particularly drawn to support? What actions might I take to offer this support?

[] Are there others in your community who might benefit from receiving this type of support? If so, how might you contribute to your community by supporting such people? Is there information that you could disseminate? A support group that you could organize? A newsletter? A means for developing new policies in your community?

[] Now ask yourself: Is there some area in my life in which I could benefit from support? What kind of support would benefit me most? Is
there someone in my social network from whom I would like to
receive support? Are there community resources that could provide
such support? Could I create a group to provide the support I and
others with a similar issue are seeking? Could I recruit an
appropriate individual to facilitate the group?

Keep in mind the many benefits of asking for, and receiving support
from others, and the benefits of giving support to others. Resolve that
you will seek opportunities both to give and to receive support.

CHAPTER NOTES

1. Frederick Hudson, *Mastering the Art of Self-Renewal* (New York: MJF Books, 1999).

2. Herbert Benson and Richard Friedman, "Harnessing the Power of the Placebo Effect and Renaming It 'Remembered Wellness,'" *Annual Review of Medicine* 47 (1996): 193–199.

3. Toshihiko Maruta et al., "Optimists Vs. Pessimists: Survival Rate Among Medical Patients Over a 30-Year Period," *Mayo Clinic Proceedings* 75, no. 2 (2000): 140–143.

4. See note 3 above.

5. Leonard A. Jason et al., "The Measurement of Wisdom: A Preliminary Effort," *Journal of Community Psychology* 29, no. 5 (2001): 585–598.

6. Thomas Berry, *The Dream of Earth* (San Francisco: Sierra Club Books, 1988).

7. Salvatore R. Maddi, Deborah M. Khoshaba, and Authur Pammenter, "The Hardy Organization: Success by Turning Change to Advantage," *Consulting Psychology Journal* 51, no. 2 (1999): 1117–1124.

8. George Anders, "How Intel Puts Innovation Inside," *Fast Company* 56 (2002): 122–124.

9. See note 8 above.

10. See note 7 above.

11. Silvia Arber, Steven J. Burden, and A. John Harris, "Patterning of Skeletal Muscle," *Current Opinion in Neurobiology* 12, no. 1 (2002): 100–103.

12. Janice M. Morse and Barbara Doberneck, "Delineating the Concept of Hope," *Journal of Nursing Scholarship* 27, no. 4 (1995): 277–285.

13. See note 12 above.

14. Trudi Venters Horton and Jan L. Wallander, "Hope and Social Support as Resilience Factors against Psychological Distress of Mothers Who Care for Children with Chronic Physical Conditions," *Rehabilitation Psychology* 46, no. 4 (2001): 382–399.

15. Pamela J. Feldman et al., "Maternal Social Support Predicts Birth Weight and Fetal Growth in Human Pregnancy," *Psychosomatic Medicine* 62, no. 5 (2000): 715–725.

16. Carolyn E. Schwartz and Rabbi Meir Sendor, "Helping Others Helps Oneself: Response Shift Effects in Peer Support," *Social Science & Medicine* 48, no. 11 (1999): 1563–1575.

17. Teresa E. Seeman et al., "Social Relationships, Social Support, and Patterns of Cognitive Aging in Healthy, High-Functioning Older Adults: MacArthur Studies of Successful Aging," *Health Psychology* 20, no. 4 (2001): 243–255.

18. Teresa E. Seeman and Eileen Crimmins, "Social Environment Effects on Health and Aging: Integrating Epidemiologic and Demographic Approaches and Perspectives," *Annals of the New York Academy of Sciences Population Health and Aging: Strengthening the Dialogue between Epidemiology and Demography* 954 (2001): 88–117.

CHAPTER 12

protect the vulnerable

ONE OF THE MOST IMPORTANT FUNCTIONS of community is protecting the vulnerable. In the community of the body, the immune system protects vulnerable organs and systems from harmful and infectious agents. In the social community, organizations such as hospices, volunteer groups, adoption agencies, and social service agencies protect vulnerable members of society—the dying, people who live with pain, children with AIDS—all of whom are in danger of being ignored, marginalized, or discarded.

Providing protection to those unable to care for themselves is as important to the long-term health of the community as protecting the brain, the nervous system, and other vital organs is to the continuing health of the body. In fact, as Nobel Peace Prize winner Elie Wiesel urges in one of his essays, "solidarity with the weak, the persecuted, the lonely, the sick, and those in despair [may be] the duty of our generation as we enter the twenty-first century."[1] In the noble and humanized community of the future, Wiesel writes, "all members will define themselves not by their own identity but by that of others."

Wiesel's remarks address what is important about our communities today and how they can thrive tomorrow. Circumstances arise that can make each of us vulnerable at some time in our lives. In such circumstances, help from others can make an enormous difference. For instance, Dave Ulrich, professor of business administration at the University of Michigan, tells of the time when he was a college freshman at Brigham Young University and he chartered a bus to transport a group of fellow students from Salt Lake City to Kansas City for Christmas vacation.[2] A snowstorm stranded the bus in Rock Springs, Wyoming. No one on the bus knew anyone in town, and none of the students had extra money for lodging or food. Ulrich took the initiative to contact an ecclesiastical leader in Rock Springs and explain the students' predicament. Soon, forty students and the bus driver had places to stay, food, and support until the roads were cleared. These benefactors taught Ulrich a lesson in community that he has not forgotten.

In this chapter we explore the messages of the body's immune system and learn how protecting the vulnerable can help us to create the humane community in which we would all like to live.

Protecting the Vulnerable Brain

To begin, let's look at the special way the body's immune system protects the vital and vulnerable brain. The soft tissues of the brain and spinal cord must be protected from assault from outside the body, as well as from disruption from within. To defend the critical tissues of the central nervous system from external injury, the brain and spinal cord are encased in bone. Ironically, the bony skull that protects the brain gives its tissues little physical room in which to expand. An inflammation that causes swelling, such as might be caused by an infection like encephalitis, an injury, or the pooling of blood within the brain following a trauma, can cause the brain to expand. As pressure develops, brain tissues strain against their bony cage. If it is not relieved, the pressure can cause neurons to die. In other words, bone, which protects the soft tissues of the brain from external trauma, can itself become an instrument of injury.

Changes in the internal environment brought about by fluctuations in concentrations of ions and molecules in the blood can also damage the brain. Isolating the brain from the ions carried by the blood requires a complex mechanism of protection, first noted more than a hundred years ago. In an experiment, a dye bound to a protein was injected into an animal's blood vessel. Soon the dye visibly stained all the tissues of the animal's body, except for those of the brain and spinal cord.[3] This experiment proved the existence of what is called the blood-brain barrier, a system of protection that keeps harmful molecules carried by the blood from permeating brain tissue.

The blood-brain barrier is critical because the transmission of impulses across neural synapses relies on the precise movement of ions, such as sodium and potassium, into and out of neurons. To prevent these transmissions from being compromised by ions carried by the blood entering brain tissues, the blood vessels that supply the brain have a special lining. Throughout the rest of the body, the cells that line the interior walls of the blood vessels have physical openings or gaps between them. Molecules exit the blood vessels by passing between these gaps. However, the cells lining the walls of blood vessels in the brain differ from those in the rest of the body in that there are no gaps or openings between them. Like the interlocking pieces of a jigsaw puzzle, these cells form tight junctions, creating a physical barrier that prevents ions and other disruptive molecules from passing from the blood vessels into the brain.

Some kinds of helpful molecules in the blood can cross the barrier into the brain because their chemical makeup allows them to dissolve into the membrane lining the blood vessels, and from there diffuse into brain tissues. For example, steroids—including estrogen, a hormone produced primarily by the ovary and vital to reproduction—easily diffuse from blood vessels into the brain. It is important that estrogen reach the brain because it plays a critical role in protecting the brain's cognitive functioning, especially as women age.

However, there is a fatal flaw in this delivery system. The chemical makeup of the glucose molecule, which supplies energy to the brain's neurons, does not allow it to pass through the walls of the blood vessels and into brain tissue by diffusion. The tight junctions of the blood-brain barrier keep glucose from moving into brain tissue by passing between physical openings between cells. Compounding the situation, neurons, which have a constant need for glucose, do not have the capacity to store glucose. Without it, they die.

How does the body make glucose available to neurons on demand while still protecting it from harmful influences? Ingeniously, a molecule called GLUT, present only on the surface of brain cells, actively ferries glucose molecules across the blood-brain barrier to meet the energy demands of neurons.[4] Other organs of the body have no need for the GLUT active transport system, as glucose moves easily through openings between cells lining the walls in the blood vessels.

The body's system of protection for the brain's vulnerable tissues is complex and unique. Protected from outside assault by the skull and from harmful internal influences by the blood-brain barrier, the brain is also continually nourished by a special transport system designed specifically to meet its needs. The brain's system of protection also suggests that if protective barriers are too rigid, like the bony skull, they can themselves be the cause of injury.

As cellular wisdom suggests, the challenge for society is to devise methods of protecting its vulnerable members—methods that are strong enough to shield them from external and internal sources of danger, yet sensitive enough to meet their need to connect with, and be nourished by, others. In addition, the method must give the vulnerable sufficient room to grow emotionally, mentally, spiritually, and physically stronger. Let's keep these criteria in mind as we look at some of the ways our communities care for their most vulnerable members.

Hospice and Protecting the Dying

Like the vulnerable brain, people who are facing death need to be connected to sources of helpful support, such as health care workers and family members. At the same time, the vulnerable need to be protected from potentially disruptive influences, both from outside sources and from within.

In the late 1940s, nurse Cicely Saunders was caring for a man dying of cancer in St. Joseph's Hospital in east London. Saunders developed a relationship with the man.[5] His experience sparked her vision of a place where patients who needed help with their pain but could not stay at home would be welcomed and cared for in a communal setting during their final weeks or months of life. After earning her medical degree, Saunders established the first hospice, called St. Christopher's, in London in 1967.

The first hospice in the United States was established in New Haven, Connecticut, in 1974. By 1998, more than three thousand hospice programs in the U.S. were caring for more than half a million people. Today, 80 percent of care provided by hospice takes place in the home of the dying person, or in a nursing home. Like the brain's system of protection, hospice care defends people facing death from both external demands and from the excessive pain and other internal consequences of the dying process. At the same time, it creates avenues for constant nourishment and support for dying people.

Hospice care works best when there is an open awareness of dying.[6] Open awareness has two components: the family members or caregivers of a person know that the person is likely to die, and the person who is dying is aware that death is immanent. When such awareness is fully open, both the dying person and the family members are able to speak openly about the coming death. Studies that have compared people dying in closed awareness with those dying in open awareness reveal that open awareness gives both the dying person and family members choices about where the death will occur. People who approach death in open awareness more often die in a hospice facility or at home, with family members present. Rushed emergency admission to the hospital is more frequent among those who maintain closed awareness. Familial support is minimized, as is the opportunity to integrate the death of a loved one into the family's history. Hospice counselors are specially trained to help families work toward open awareness, while protecting people who are dying from well-meaning, but sometimes intrusive interactions with loved ones.

I experienced the protective barrier hospice creates to safeguard the rights of the dying when my own mother was dying of cancer. Because I felt compelled to help her get well, I urged my mother to eat, despite her refusals. Then a hospice caregiver explained that my mother's choosing not to eat was a sign that she was disengaging and beginning to die. Hospice workers visited my mother every day during this time. They advised me to honor my mother's dying process and allow her to set the pace.

Though my mother had stopped taking all medication and spoke with great difficulty, the calm atmosphere hospice care provided created the opportunity for my mother to speak with me about arranging her financial and legal affairs, and reviewing her instructions regarding her funeral.

As my mother's death approached, hospice workers helped me to recognize the changes in her breathing that heralded the last stages in her dying process. With this knowledge, I chose to stay close to her that day, releasing her bit by bit. That last day, it was especially difficult for my mother to speak. Though I had known that she wanted to be cremated, her final words shocked me.

Deliberately, she removed the oxygen mask and struggled to enunciate each word. "How late are they burning?" she asked me.

Tears welled up, and I couldn't answer. Turning to my husband, I repeated her words and asked, "She couldn't mean the crematorium, could she?"

He nodded.

Forty-five minutes after asking this question, my mother died.

Without hospice, these last precious moments with my mother might have been lost to me. Her words let me know that she had accepted her death. Hospice care allowed me to share her last moments of surrender. Without hospice, my mother's death would have been surrounded by anxiety, not by gentleness and love.

Like the bony skull that protects the brain from assault from outside, hospice created a shell of gentleness and respect around my mother. Like the blood-brain barrier, it allowed in only those influences that would be helpful to her internal process, giving my mother the freedom to die in her own way.

Protection from Pain

In addition to the external protection hospice provides, people facing death need to be protected from excruciating or incapacitating pain. Oregon Senator Ron Wyden has taken the lead in fighting for the rights of Americans to be free of pain. According to Senator Wyden, at least fifty million Americans suffer from chronic pain.[7] Often, the pain of older patients is untreated or under-treated because the health care system tends to focus on the cure of disease rather than on the management of the patient's pain or symptoms. Moreover, Wyden says, 40 to 50 percent of patients experience "moderate to severe pain at least half the time in their last days of life."

To redress this public health problem, Wyden is sponsoring a bill in the Senate called the Conquering Pain Act,[8] which seeks federal help to improve the quality of life for people who are dying. If passed, the Conquering Pain Act would provide federal funds for education, websites, research, demonstration projects, and community resources, such as regional family support networks for pain and symptom management. As of this writing, Wyden's bill is still in committee. However, Wyden's focus on managing pain through hospice care has not gone unnoticed in his home state of Oregon. Statistics show that 30 percent of Oregonians who died in 1997 did so in hospices, more than 10 percent above than the national average.

Relieving the pain of the dying was at the heart of Cicely Saunders's dream of hospice service. To her, pain had multiple dimensions: physical, psychological, social, emotional, and spiritual. Saunders understood that pain is more than a bodily sensation, yet the successful control of physical pain helps people cope with pain's other dimensions.

The body experiences two types of pain. One type, called epicritic pain, is sudden and sharp, such as we experience when the body receives a physical insult—for example, when a needle pierces the skin. We experience epicritic pain quickly because only three neurons are involved in transmitting the information from the spot of the insult to the cerebral cortex: one neuron from the spot where the needle pierces the skin into the spinal cord; a second from the spinal cord to the thalamus, a relay station deep in the brain; and a third from the thalamus into the cerebral cortex. When the sensation registers in the cortex, we experience localized pain.

A second type of pain, called protopathic, often accompanies the sharp pain described above, but is experienced after a short delay. This second type lasts longer, is more diffuse—for example, it may be experienced as a burning sensation—and is more difficult to localize. A host of neurons in the central core of the spinal cord carry the impulses of protopathic sensations. The neural information is relayed to another part of the thalamus, and from there, it projects to almost all regions of the cortex. As a result, we experience dull, enduring pain.

The control of pain—from injuries that healthy people experience, as well as from diseases that many people facing death experience—depends on blocking the transmission of neural impulses. A set of descending pathways running from the cortex to the spinal cord parallel the ascending pathways discussed above. Activating the descending pathways can interrupt and block the transmission of pain impulses. The mechanism for activating the descending pathways involves the body's own opiates, a class of chemicals called endorphins, which block the transfer of pain information from the site of an injury to the spinal neuron. This blocking process happens naturally in many circumstances. For instance, professional athletes are often able to play through an injury because natural endorphins block pain receptors.

The drug morphine, prescribed for many dying people to control their pain, works on the same principle. The drug binds to endorphin receptors at several sites in the body. At one site in the brain stem, activated endorphin receptors trigger the descending pathway to block pain transmission. At a second site in the spinal cord, activated endorphin receptors block transmission of pain from spinal cord neurons. If neurons in the spinal cord do not transmit impulses to the thalamus, no pain is experienced. Activating endorphin receptors through the administration of morphine can block both epicritic and protopathic pain pathways.

A recent study was conducted to develop guidelines for administering morphine and other pain blocking medications to dying patients.[9] Scientists looked at the effects of an integrated program of pain management on 168 patients who died over a twelve-month period at an inpatient hospice at the Royal Liverpool University Hospital. In addition to pain-blocking medications, patients at the center receive emotional and psychological support; physical therapy, including complementary therapies; and spiritual care. Multidisciplinary teams of staff operate not only within the center but also in the community, linking local hospitals, general care physicians, and volunteer supporters.

A hallmark of the integrated care management approach is that that dying patients are prescribed medication as required for pain and agitation, to prevent delays in the control of symptoms, avoid unnecessary distress for patients and relatives, and—in some circumstances—even prevent hospital admission. The nursing staff in the hospice monitors symptoms of pain and agitation every four hours in an ongoing assessment of patient needs. Using this protocol, a majority of patients in the study, 80 percent, experienced no or only one episode of pain or agitation during the two days prior to their deaths. Within eight hours of death, results were even better, with less than 4 percent of patients experiencing episodes of pain. The integrated pain management approach seems well in keeping with the holistic goals of the hospice movement, as expressed by Cicely Saunders: "You matter because you are you. You matter to the last moment of your life, and we will do all we can, not only to help you die peacefully, but to live until you die."

While this philosophy is key to the hospice movement, it seems clear that physicians could do much more to assist their dying patients in relieving their physical and emotional pain. End-of-life care provides many opportunities for physicians to provide emotional support to dying patients and their families, such as encouraging them to use the services of home hospice. The dying may want the opportunity to resolve unfinished business, remember personal accomplishments, and say goodbye to important people, a process sometimes referred to as integrated dying.[10] Interestingly, physicians are less likely than patients to credit the importance of feeling that life is complete.

What can be done to encourage physicians to do more to protect patients from pain and suffering at the end of their lives? First, the medical model must alter so that end-of-life care is valued as much as curative care. Second, physicians must be trained to deal with the relief of pain, dying, and bereavement. A program at the Wellington School of Medicine in New Zealand points the way.[11] Student doctors who care for hospice patients meet and interview each patient's family at least twice. Students also create a portfolio of notes about these visits, with observations about any emotional and ethical issues relevant to the patient, in addition to the patient's medical history. Besides meeting with family members and chronicling the patient's medical condition, the students write essays about this aspect of their training and meet to discuss their experiences with groups of their peers.

Finally, physicians must be taught to value caring and intimacy, which often arises during end-of-life situations. One doctor recounted his experiences in caring for a dying man named Roy in a scientific journal article by Dr. Roderick MacLeod.[12] The doctor describes how he came to understand the human dimension of caring for the dying:

> I was a GP and visited [Roy] daily for weeks. I could still describe the room, the medications, the dog, the care, and even his clothes! He and his wife had been married for decades, but they also worked with each other daily. He taught me about symptoms, and he taught me the importance of making accurate assessments, because I got it wrong sometimes. They taught me of the importance of broadening the medical role to include an acknowledgement that in many ways I was powerless and vulnerable myself. Towards the end of his life there were no truly "medical" things to do. . . . I was their companion; there was no curative or "medical" role in the accepted sense. Despite knowing that he was dying, when it came, I was devastated.

With proper training, and working in tandem with nurses and hospice caregivers, it seems clear that physicians can do much to protect the dying. New tools such as integrated care management may provide guidelines for physicians and other caregivers that guarantee that the dying can be free of pain, maintain their dignity, and make the choices that allow them to complete their lives as they have lived them—with choice and support.

The Immune System and AIDS

The body's way of protecting the brain, as we have seen, has features that meet its unique needs. Throughout the rest of the body, an equally comprehensive system of protection guards other organs and systems from potentially harmful outside agents, including pathogenic bacteria and viruses. These invaders are carried by the air we breathe, the food we eat, the water we drink, and the substances with which the skin and other surface-contacting membranes come into contact. To protect the body's health, the immune system seeks out and destroys most potentially disease-causing agents. The key warriors in this battle are the lymphocytes, a type of white blood cell.

Initially, lymphocytes are born in the bone marrow. Some lymphocytes remain in the bone marrow until they mature as B cells; others migrate to the thymus gland, a butterfly-shaped organ at the base of the neck, where they mature as T cells. Early in their development, B and T cells produce specific receptors that reside on their exterior surfaces. These receptors are capable of recognizing most of the invading agents we may encounter during our lives. About one trillion receptors are generated in this process on each B or T cell.

T cells circulate though the body carried by a fluid called lymph. Lymph is found in many sites throughout the body, especially in the lymph nodes—small, bean-shaped structures in the neck, the pit of the arm, and the groin—as well as in the spleen, the tonsils and adenoids, the appendix, the intestines, and in tissues of the gut. When T cells encounter a potentially disease-causing agent, or antigen, in one of these locations, they alert B cells to produce antibodies. Antibodies are Y shaped protein molecules that attach to antigens. The antibodies disable the antigen and signal for T cells to destroy the invading cell. In response to this signal, T cells move to the thymus gland where they divide and produce clones of themselves to combat the foreign substance. Varieties of T cells—repressor T cells, helper T cells, and killer T cells, each with different receptors—work together to mobilize the body's immune response. T cells enter the blood stream to reach sites throughout the body and destroy the antigen-bearing foreign cells. Unlike B cells, killer T cells do not secrete antibodies. Rather, they recognize an invading cell and bind to it. Then they secrete a burst of chemicals that open holes in the invading cell's membrane. The invading cell literally bursts open and dies.

The immune system itself, however, can come under attack. A most potent enemy today is the human immunodeficiency virus (HIV), the antigen that causes acquired immune deficiency syndrome (AIDS). In 2001, the Centers for Disease Control and Prevention (CDC) estimated that three million people died from AIDS worldwide and that forty million people were infected with HIV. In the United States alone that year, the CDC reported more than 450,000 deaths from AIDS and almost 800,000 active cases.

The devastating medical and social consequences of AIDS are well-known to many people, though fewer may understand the physiological mechanism through which the disease functions. Briefly, by attacking the body's immune system, AIDS makes sufferers vulnerable to other diseases, such as pneumonia, cancer, and neurological disorders. It is these conditions, rather than the AIDS virus itself, that cause so many people to die of the disease.

HIV is a potent infectant of cells. Once it enters a cell, HIV can remain dormant for a long time. When it is activated, however, the virus takes over the cell's functions and uses them to produce more and more replicates of itself. A single infected cell can produce thousands of copies of HIV.[13] Some of these copies enter the lymph nodes. There, a specific variety of T cells, called T4 cells because of a receptor on its surface, becomes the target of the HIV infection. In simple terms, in response to the presence of the virus, T4 cells move into the lymph nodes and divide to form more T4 cells. HIV reproduces variants or mutants of itself, each of which activates a response from different T4 cells. This repeating process results in chronic immune activation, which destroys the effectiveness of the T cells to detect and combat other infections. Moreover, the presence of HIV disrupts the generation of new lymphocytes in the bone marrow and keeps lymphocytes from maturing into T cells in the thymus gland. Without T cells, the immune system is incapacitated and eventually breaks down completely.

HIV is transmitted from one person to another by the exchange of fluids, such as through sexual contact, the exchange of blood, or sometimes between mother and newborn infant. Co-infection with other antigens, such as herpes simplex, hepatitis B, tuberculosis, and sexually transmitted diseases, speeds up the depletion of T4 cells. The immune system can be further impaired by the use of hard drugs and by mental or physical stress. The current prediction is that HIV will continue to be a potent virus because of its fast mutation rate, its ability to recombine and form new strains, and its ability to use the human immune system to replicate itself.

Think for a moment about AIDS in the light of cellular wisdom. What dangers as difficult to combat as AIDS threaten the body of society? Perhaps such evils as racial and religious intolerance, poverty and hunger, and the abuse of women and children come to mind. Like AIDS, these threats are spread from person to person, including from parents to children. Ask yourself, What forces can communities muster to combat these dangers? Most important, ask, What can I do personally to defend the community in which I live from such devastating and potentially deadly invaders?

Protecting Children with AIDS

Because the AIDS virus is most usually transmitted by sexual contact with an infected person, AIDS is a potent killer of adults. Yet children, too, can be victims of the disease. According to the World Health Organization, 3.2 million children under the age of fifteen worldwide are living with AIDS. In 2002 more than half a million children under the age of fifteen died of AIDS.[14] In the United States, nine thousand children under the age of thirteen were diagnosed with AIDS cumulatively through 2001, and more than five thousand children died.[15]

More than 90 percent of children who contract AIDS do so because they are born to an HIV-infected mother. The mother can transmit the virus to her baby during delivery through the vaginal canal or by breast-feeding. HIV can also cross the placenta and infect the liver, bone marrow, or thymus of the fetus, the organs in which the fetus generates T cells. The best protection for the fetus is treating an HIV-infected mother during pregnancy with a cocktail of drugs that attack the ability of HIV to mature and reproduce.

Until we are able to cure AIDS, the community has a special responsibility to protect its most vulnerable members from the physical, emotional, and social consequences of this devastating disease. Caring for children who face an uncertain future because of AIDS differs in several important ways from caring for adults who are coping with the disease. Children often do not understand why they are suffering the fatigue, pain, and discomfort of illness. The unpredictability of their physical functioning generates distress and feelings of fear of the unknown, pain, suffering, and loss of friends and family.

Sick children often display mood swings, including anger and temper tantrums. In a 1993 study, children with AIDS were asked to complete the sentence, "I often wonder. . . ." One child's heartbreaking response was: "I often wonder how often I'm going to get sick and what will happen to me during those times. . . . I wonder how much longer in life I have. Sometimes I think I only have months to live, other times I'm more hopeful and I think I'll live at least a couple of more years. The thought of not living long scares me. Especially dying."[16]

Children facing death also differ from adults in that they do not understand what death is. A child's conception of death changes as she matures. Children ages three to six often view death as an immobilization that is reversible. They may fear that their own thoughts can cause death. The fact that death is not reversible is recognized by children ages seven to twelve. They also understand that death happens to all living things. Teenagers develop a stronger reality-based notion of illness and its relationship to death. Fully aware of the finality of their approaching death, ill teenagers, like their healthy peers, have a growing need for supportive connections with people their own age outside of the family. Parents of a child with AIDS often fear talking to the child about the condition, because of the effect the news may have and their inability to provide reassurance and comfort. They also fear the social ostracism that might disconnect the child from his circle of friends.

Parents face difficult decisions about hospice care. All evidence shows that home hospice provides the most nourishing environment for a dying child. Yet, parents may feel that they cannot provide appropriate care for acute medical conditions, such as infections. Managing pain also becomes a prime concern. Children with AIDS need a regular program of pain medication.

"Children who die young are some of our greatest teachers," said death and dying expert Elisabeth Kubler-Ross during an interview.[17] In her book *AIDS: The Ultimate Challenge*,[18] Kubler-Ross recounts her experiences in trying to establish a hospice for children with AIDS on her two hundred-acre farm in Virginia in 1985, when fear about AIDS was at its most intense. Opposition to the AIDS hospice began immediately. The transcript of a public meeting called to discuss the hospice plan reveals how little understanding existed at the time regarding AIDS. The people of Highland County, Virginia, wanted a guarantee from Kubler-Ross that she would not "pollute our water system, our sewer system with AIDS." Because of this level of fear and misunderstanding, Kubler-Ross failed to get the rezoning required to build the hospice.

This setback did not deter Kubler-Ross from her efforts to protect children with AIDS. In 1986, she began lining up foster parents in her community to care for children with AIDS. This plan, too, drew angry opposition. When she returned from a trip, friends warned her not to return home but to stay with them instead. Ironically, she had no home to which she could return. A fire had destroyed everything, including her records of research with dying people, her manuscripts, and years of accumulated personal belongings, pictures, and family remembrances.

But Kubler-Ross had ignited a movement of protection that no fire or opposition could stop. Many couples soon stepped forward and volunteered to adopt HIV-positive children. One such couple, Joy and Jim Jenkins, adopted an HIV-positive infant not knowing whether or not the baby would develop AIDS.[19] However, they did know that if the antibodies present in their adopted son's body were residuals from his HIV-positive mother, they would disappear within the first year. If the antibodies were still present after that year, they would know that the child had developed AIDS.

Joy and Jim named their new son James Michael and prepared to give the baby a year of loving care and protection, despite the uncertainty about his future. While they waited for the baby to come to their home from the hospital where Joy Jenkins worked as a nurse, the couple took a special course in foster parent training. During the course, an AIDS educator asked them if they had faced the possibility that the baby might die. The question shocked them into profound awareness of what the future could hold.

That first year was a roller coaster of hopes and fears. One night, James Michael's fever soared to 104 degrees. Joy feared the worst. In the midst of her worrying, she became nauseous and found herself rejoicing that she was sick, too. As it turned out, both she and the baby had contracted food poisoning from a frozen custard they had shared. After nine months, James Michael still tested positive for HIV. Miraculously, at fifteen months, HIV antibodies were no longer detectable in his blood. The Jenkins were overjoyed!

Wanting to do more for AIDS throw-away babies, Joy and Jim founded the Children with AIDS project, which matches families with adoptable HIV-positive children and provides post-adoption emotional support. At the time of this writing, the project has matched nine couples with AIDS babies. Among the adoptive couples are the Jenkins themselves, who adopted another HIV-positive child, a little girl they named Arlis.

Joy and Jim Jenkins's active support for AIDS children and their families turned the vision of providing protection for society's most vulnerable members into a viable community program. Barriers constructed of limiting beliefs did not stop them. The question, What can a couple of people do about a worldwide epidemic? did not stop them. Rather, they took the personal step of adopting an at-risk child themselves, and they allowed their example to inspire other adoptive families. Like the body's immune system, the Jenkins's personal commitment to fighting ignorance, prejudice, and fear demonstrates what can be done to combat the forces that threaten the body of society. It is actions like theirs that will help us make the noble and humanized community of the future a living reality.

PROTECTING THE VULNERABLE

Part I: Exploring Your Attitudes

Find a quiet, comfortable place where you can settle in for a while without being disturbed. Reflect briefly on the issues discussed in this chapter. In this first part of this exercise, we explore the attitudes that you hold about these topics.

When you are relaxed and centered, think back over the experiences with vulnerability, pain, illness, or death that you or your loved ones have faced. Now ask yourself:

[] What vulnerable people have you taken responsibility for protecting in your life? What were the circumstances that necessitated this protection? What protective actions did you take? Is there anything you might now do differently?

[] Have you or anyone you know been faced with persistent or chronic physical pain? What were the circumstances? What steps did you take to help them get relief of the pain or comfort them, if relief was impossible?

[] If you have not had to deal personally with pain, how do you feel about the possibility of having to do so? What is your attitude about you or a loved one taking pain medication, should the circumstance arise?

[] What circumstances in your life, if any, have caused persistent or chronic emotional, psychological, or spiritual pain? What steps have you taken to relieve this pain? What more might you do?

[] Recall an experience of dealing with the death or impending death of a friend or family member. Did the death take place in open or closed awareness? What discussions, if any, did you have with the person facing death? Where did the death take place? Who was present?

Was hospice care involved, and if so, what was your experience with hospice?

[] What steps did you take to alleviate the dying person's physical, emotional, or spiritual pain? Is there any circumstance surrounding the death that you would now approach differently?

[] Recall any experiences you have had in dealing with AIDS. How would you describe your attitude toward the illness? Has your attitude changed over time? If so, in what ways?

[] How would you describe your attitude toward your own death? Would you say you are relaxed and at ease? Fearful? Or would you rather not think about it? What steps might you take to overcome your fears and cultivate a more relaxed and accepting attitude?

[] After reading this chapter, what special instructions might you wish to leave for your loved ones about how you wish to be cared for in your final days? If it seems appropriate, consider whether you wish to put these instructions in writing or add them to a living will.

Part II: Protecting the Vulnerable in Your Community
In the second part of this exercise, you are invited to explore what your role might be in protecting the vulnerable members of your community.

[] Thinking of your particular community, list the groups that seem to you to be particularly vulnerable; for instance, young children at home alone after school while their parents work; teenagers with no safe place to gather; immigrants who need help adjusting to American life; people who suffer from cancer, AIDS, or some other life-threatening illness; residents of homes for the elderly or nursing homes; the mentally ill.

[] What programs are in place in your community to protect vulnerable groups? What might you do to assist these programs as a volunteer or contributor?

[] What additional programs do you see as needed? What steps might you take, alone or in conjunction with your neighbors or with members of your church or other community organization, to extend protection to a vulnerable group in your community?

[] What legislation do you see as needed to protect the rights of some vulnerable group in your community, state, or the country at large? The Internet is a wonderful resource for researching legislation. Take the time to reach out electronically to learn more about what is being done to protect the vulnerable, and to consider what help you might offer.

As a concluding step, share your desire to protect the vulnerable with at least one other person. Develop a plan with your friend. Identify one next step you'd like to take. Ask your friend for whatever support you might need to take this step. Set a date by which you plan to take action. Make a specific time and date to meet with your friend to discuss the outcome of your action.

]OOO[

THE BODY TEACHES US that vigilance in protecting its vulnerable organs and systems is a vital and dynamic aspect of life. In society, as well, the vulnerable are an integral element within our communities. When we ignore, deny, marginalize, or discard the vulnerable, we ignore a precious opportunity. The vulnerable have the potential to evoke our caring. They call us to live in community fully, as people with a sustained carefulness for all members. When we answer the call to protect the vulnerable, we and our communities thrive.

CHAPTER NOTES

1. Elie Wiesel, "Afterword," in *The Community of the Future,* ed. Francis Hasselbein et al., Drucker Foundation Future Series (San Francisco: Jossey-Bass Publishers, 1998), 273–275.

2. Dave Ulrich, "Six Practices for Creating Communities of Value Not Proximity," in *The Community of the Future,* ed. Francis Hasselbein et al., Drucker Foundation Future Series (San Francisco: Jossey-Bass Publishers, 1998), 155–166.

3. L. L. Rubin and J. M. Staddon, "The Cell Biology of the Blood-Brain Barrier," *Annual Review of Neuroscience* 22, no. 1 (1999): 11–28.

4. Mark S. McAllister et al., "Mechanisms of Glucose Transport at the Blood-Brain Barrier: An In Vitro Study," *Brain Research* 904, no. 1 (2001): 20–30.

5. David Clark, "'Total Pain,' Disciplinary Power and the Body in the Work of Cicely Saunders, 1958–1967," *Social Science & Medicine* 49, no. 6 (1999): 727–736.

6. Clive Seale, Julia Addington-Hall, and Mark McCarthy, "Awareness of Dying: Prevalence, Causes and Consequences," *Social Science & Medicine* 45, no. 3 (1997): 477–484.

7. Ron Wyden, "Steps to Improve the Quality of Life for People Who Are Dying," *Psychology, Public Policy, & Law* 6, no. 2 (2000): 575–581.

8. *Conquering Pain Act of 2003*, S 1278, 2003.

9. John Ellershaw et al., "Care of the Dying: Setting Standards for Symptom Control in the Last 48 Hours of Life," *Journal of Pain and Symptom Management* 21, no. 1 (2001): 12–17.

10. Karen E. Steinhauser et al., "Preparing for the End of Life: Preferences of Patients, Families, Physicians, and Other Care Providers," *Journal of Pain and Symptom Management* 22, no. 3 (2001): 727–737.

11. Roderick D. MacLeod, "On Reflection: Doctors Learning to Care for People Who Are Dying," *Social Science & Medicine* 52, no. 11 (2001): 1719–1727.

12. See note 11 above.

13. Janis Faye Hutchinson, "The Biology and Evolution of HIV," *Annual Review of Anthropology* 30, no. 1 (2001): 85–108.

14. World Health Organization, "AIDS Epidemic Update" (2002).

15. Center for Disease Control, "HIV Surveillance Report: U.S. HIV and AIDS Cases Reported through December 2001" (2001).

16. Robert W. Buckingham and Edward A. Meister, "Hospice Care for the Child with AIDS," *The Social Science Journal* 38, no. 3 (2001): 461–467.

17. Elizabeth Kubler-Ross, in conversation with John Harricharan, http://www.insight2000.com/kubbler-ross.html (accessed September 23, 2003).

18. Elisabeth Kubler-Ross, *AIDS: The Ultimate Challenge* (New York: Touchstone, 1987).

19. Joy and Jim Jenkins' story was told in *Good Housekeeping* magazine and is reprinted on the website www.aidskids.org.

CHAPTER 13

the richness of diversity

MUCH OF THE CONFLICT THAT THREATENS the world today has its roots in differences between people. Differences in nationality, race, ethnicity, religious belief, gender, and sexual orientation, which many celebrate as the richness of diversity, are also the cause of much strife and discord. Yet, judging from the way conflict functions in the biological world, diversity and the friction that it engenders may be a necessary catalyst to all evolutionary movement, creativity, and growth. In the cellular world, evolutionary biologists tell us conflict stimulates cells to devise adaptive strategies to meet and regulate it. In fact, without conflict, the evolutionary shift from individual cells to multicellular organisms—the shift that made possible the evolution of the biological world we know, including human life—would not have occurred. In the cellular world, cooperation emerged from conflict.

In this chapter, we explore diversity and look at ways that social conflict can spur us to evolve new strategies for cooperation. We examine how cultural differences impact our perspectives, and how bias or prejudice threatens the life of a community. Taking our hopeful cue from the cellular universe, we see how transcending bias and or prejudice, which are limiting and separating concepts can herald a new evolutionary phase in social consciousness and help us create diverse, but harmonious human communities.

Cellular Conflict as a Spur to Cooperation

The shift from life as a single cell to membership in a coherent complex of cells stands out as a major event in evolution. If this leap had not occurred, the plants and animals we know today would not exist. What made this step possible?

To function independently, a cell needs to be able to engage in all the activities required to sustain life, including the ability to reproduce. As cells came together into multicellular aggregates far back in our evolutionary history, they necessarily gave up some of these abilities in return for new capacities available only to multicellular organisms. For instance, larger organisms can fend off predators more effectively than single cells. They can also generate their own internal environment, giving cells that aggregate into larger organisms more control over changes in external conditions.

Scientists now believe that these new capacities came at a price.[1] As cells joined together, they transformed, such that different groups of cells specialized in different activities. For instance, in aggregates, cells could not continue to reproduce at a level equal to that of independent entities, and many, not at all. Particular cells evolved so as to be devoted exclusively to the process of reproduction, like the ova and sperm. This specialization of reproductive capacity benefited the aggregate as a whole. Nonreproductive cells became smaller and more functional, and the dangers of unrestricted reproduction—like the cellular mutation that leads to cancer—were minimized.

Freed of the need to replicate themselves, nonreproductive cells differentiated into distinct cell types, such as bone cells, muscle cells, liver cells, glandular cells, and neurons. As we have noted in previous chapters, cells in a multicellular organism cannot operate independently. Cells and systems must cooperate in order for us to walk, sleep, breathe, read poetry, paint paintings, compose music, understand love, and help others. Even neurons cannot function alone. The roughly one hundred billion neurons in the adult brain cannot do what they do to make us human without supporting cells that are not neurons, which outnumber them ten to one.

As specialized cell types joined together into cooperating networks, organisms became more stable and sophisticated. We have looked at many examples of behaviors made possible by such cooperation. Walking, for instance, requires that neurons act in concert with muscle cells and that muscle cells act in concert with bone cells. For such cooperation to be effective, each cell must be true to its type, and groups of specialized cells must work together. Biologists call such working together *integration*. In an integrated organism, each cell type augments its contribution to the organism by integrating its function with those of other cell types, tissues, and systems. As a result of this integration, conflict is minimized, and cooperation allows the organism to develop new and more advanced behaviors.

Conflict and Cooperation in Human Communities

As in the cellular world, human beings give up the right to complete independence to reap the benefits of living together cooperatively in community. Sociologists cite many examples of such social cooperation. In a healthy society, as in a healthy body, groups of individuals band together to benefit all members of the group. For example, in an ethnic niche, a particular group colonizes a sector of employment so that members of the group have privileged access to job openings.[2] There are such niches in restaurant work, garment factories, police and fire departments, and branches of New York's and Miami's civil service, among others. Group members find jobs for others in the group, teach them skills, and supervise their performance.

While social groupings facilitate cooperation, they can also lead to conflict. A classic study of social conflict and cooperation among adolescent boys offers clues as to how group identification can lead to conflict and, more important, how such conflict can be transformed into cooperation. In the Robbers Cave study (so-called because the study took place in Robbers Cave State Park in Oklahoma, where train robber Jesse James was said to have had a hideout), twenty-two boys spent three weeks at a camp during the summer of 1954.[3] The twelve-year-olds arrived separately and were assigned to one of two groups, the Eagles and the Rattlers.

For the first week, the groups were kept apart. During the second week, the groups were brought together for a series of athletic competitions, such as tug-of-war, baseball, and touch football. These contests led to escalating hostility between the groups, evidenced by the boys raiding each other's cabins, destroying and stealing property, and collecting sticks, baseball bats, and socks filled with rocks for use as weapons. Fistfights broke out, and food fights erupted in the dining hall.

During the third week, the psychologists running the study arranged for intergroup contact under noncompetitive conditions. Initially, this contact did not calm the hostilities. The psychologists then introduced a series of goals that could be achieved only by cooperation between the groups. For example, after sabotaging the camp's water supply, the staff announced that there was a leak in a pipe between the reservoir and the camp. Both groups volunteered to help locate the leak. The Eagles and the Rattlers were assigned to separate search parties. To find the leak, however, the Eagle and Rattler parties had to work together. When the boys found the leak, there was general rejoicing. The Rattlers even let the Eagles get ahead of them in the line to get drinks because the Eagles had no canteens with them. After drinking, the boys drifted away and mingled together. This event constituted the first friendly interaction between the two groups.

In a similar ploy, the boys were transported to a nearby lake for an overnight camping trip. Soon after they arrived, it was time for lunch. However, the truck that was to bring the food would not start and was facing uphill. One boy suggested that they get "our" tug-of-war rope and have a joint tug-of-war against the truck, proclaiming that working together, the two groups could pull the truck uphill. When they succeeded, both Eagles and Rattlers cried, "We won the tug-of-war against the truck!" Much mingling occurred after this event, including friendly talk and back slapping.

When the boys were categorized into groups, differences between boys belonging to the same group were minimized, even ignored. As a consequence, boys in the same group appeared to each other to be more similar than they actually were. At the same time, inclusion in one group enhanced differences between that group and the boys in the other group. These perceived differences led to conflict and hostility. [4]

Three strategies characteristic of group dynamics helped to transform this conflict into cooperation. [5] Personalized interactions allowed the boys to come to know each other in ways other than as members of the group. This contact de-categorized the boys from their groups and re-categorized them at a level beyond that of belonging to one of the groups. In the process, the boys become aware of qualities that cut across group boundaries. Finally, mutual differentiation encouraged the boys to emphasize their individuality, so that functions could be divided in ways that emphasized the strengths of each group in the context of cooperative interdependence.

These sociological examples demonstrate a strong parallel between the dynamics operating in the body and those that shape our communal interactions. In both cases, what shifts groups from conflict to cooperation is a perceived common need. When there is work to be done to benefit all, boundaries between groups dissolve and cooperation replaces conflict. We

need look no further for an example of this phenomenon than the cooperation among New Yorkers in the hours and days following the attack on the
World Trade Center. When the immediate need for cooperation ended, ethnic
groups reasserted their individuality, battling, for example, over the racial
makeup of firefighters to be depicted in a proposed memorial statue.

Commonality of Descent as an Aid to Integration

Though the Eagles and the Rattlers perceived themselves to be quite different from each other at the outset of the Robbers Cave study, all the boys at
camp were, in fact, the same age, and they all came from Oklahoma City. In
other words, the boys in the two groups were more alike than they were different. The same is true among the various specialized cell types in the body.
Every cell in a multicellular organism is linked to every other cell because all
arise from a single reproductive event. This commonality of descent means
that all the cells in an organism share the same DNA. Cells differ only in
which genes are turned on and which molecules are generated as a result.

For instance, two decades ago, neuroscientists realized that brain pep-
tides are produced in the gut of adult humans, as well as in the brain. Given
commonality of descent, this finding is not so remarkable. All cells have the
same DNA or potential capacity. When genes activate the same capacity in
cells in different places in the body, the same molecules are generated, leading to the differentiation of identical cell types. Thus, the shared DNA
among all the cells of a body facilitates cooperation and makes integration
between specialized cell types and systems possible. As we have seen, reproduction requires the integration of specialized cells in the brain, pituitary,
and sexual organs. The same is true for many other body processes. Specialized cells sharing common DNA participate in dynamic interactions as
required by the body as a whole.

Could having a common ancestor also be a basis for human cooperation?
The goal of The Human Genome Diversity Project, first proposed in 1991,
was to discover whether human beings share common ancestry.[6] The study
aimed to collect DNA from populations throughout the world for study by an
international group of scientists, including geneticists, physical anthropolo-
gists, paleontologists, and archeologists. While various political and logistical
delays may prevent the completion of the study, unravelling man's ancestry in
detail will prove useful in understanding our past and our future.

Individual studies similar to those planned as part of the Genome project are using variations in DNA obtained from human fossils to explore the origins of the human species. These data suggest that modern humans lived in eastern and southern Africa more than one hundred thousand years ago; in China, sixty thousand years ago; in Australia, fifty thousand years ago; in Europe, thirty-five thousand years ago; and in northern Asia, thirty thousand years ago.[7] Some scientists speculate that modern man originated in Africa and migrated to other continents. Consistent with this speculation is the relatively slight genetic variation found in African populations compared to the broader genetic variation in non-African populations. As people migrated and became separated by deserts, mountain ranges, and oceans, they became more different genetically. Data suggest that the non-African populations arose from the African populations, and that all human beings are descended from a very small group of people.

Just as shared DNA provides the foundation for dynamic group interactions in the body, the understanding that all humans share a common background of genetic information could provide the basis for more harmonious human social interactions. Author Henry T. Greely notes that as we humans are of "one family, genealogically connected, quite literally, as cousins—some first cousins, some 900th cousins, but all related," we might more readily resolve conflicts by using strategies for getting to know each other outside our national and ethnic groups.[8] An innovative educational program with this aim is discussed at the end of this chapter.

Conflict and Prejudice Between Races

If all human beings are cousins, as work on the Human Genome Diversity Project suggests, how and why does racial prejudice arise? Recent studies underscore the view that racial conflict is a learned behavior, based on culturally conditioned bias. Though stereotypes conditioned by culture sometimes trigger our responses to people who are different from us, when we make a conscious effort, we do have the ability to transcend our learned bias and to think and act in new ways.

Sociologist Paul Di Maggio explains how bias operates: Faced with a complex world, human beings partition what they encounter into manageable chunks, or categories, of related features. These categories generate mental models called *schemata*. Schemata filter data quickly and efficiently.

Hidden from view, they operate implicitly and are not verbalized. When we meet someone who has observable characteristics that fits one of our schemata—skin color, gender, sexual preference, country of origin—we apply these hidden mental models and attribute to the unknown person traits, values, and attitudes characteristic of this group. Further, we anticipate that the person will demonstrate a particular range of behaviors. While such mental shorthand is convenient, schemata often mislead us.[9]

The contrasting method of processing information, a mental process that is explicit and verbalized, requires time and attention. This mode is slow and deliberate. If we meet someone who would normally evoke our schemata, but who in some way does not conform to our mental map, we may recognize that our automatic mode of processing is inadequate and shift to a deliberate mode. Shifting from an automatic mode to a more deliberate way of processing gives us the opportunity to articulate our perspective explicitly to ourselves, and to verbalize it. Both ways of processing information bring our schemata to consciousness and help us formulate a new view.

I have experienced how schemata operate. Many African Americans lived in our neighborhood while I was growing up in New Orleans. However, the only black person I knew well was Ella, who took care of me while my parents worked. Based on my experience with Ella, I generated the schemata that black people take care of other people's homes and families. Ella let me know that she could handle anything I could throw at her, including my stubbornness. When my behavior got out of line, Ella would put her hand on her right hip, slouch to the right, throw back her head, and squint at me.

I remember once when I got a chair and a hammer and said, "Ella, you come here. I want to knock some sense into you." Taking her characteristic pose, hand on hip, Ella cocked her head, squinted at me, and with a gentle smirk said, "Miss Joan, give me that and get down!" There was no disobeying Ella. Though I loved her fiercely, she was the only black person I knew well. I would play with a friend a block away and return to our home, behind the grocery store, walking through a area of homes where many black people lived. I did not know them. I knew and loved Ella, but I was afraid of black people I did not know. I sensed that Ella might have had a tough life and may have been hurt by men. The men in our neighborhood appeared strong. I was afraid that they might hurt me, too.

Years later, in my twenties, an opportunity arose which allowed me to revise my schemata. I regret that I had few opportunities to interact with black people sooner. My first excursion into the world occurred while I was still a nun and a college chemistry teacher. I was studying advanced chemistry and physics in a summer program at American University in Washington, D.C., and doing research in physical biology at the National Institutes of Health (NIH). One day, wearing my nun's habit, I entered the NIH cafeteria for lunch. As there were no empty tables, I approached a black woman sitting alone and asked if I could join her. I recall vividly that my knees were knocking. I felt awkward and ill at ease; I thought she might rebuke me or ask me not to sit there. As we ate, we chatted. I had never spoken to a black person other than Ella and did not know how to behave or what to expect. The woman's name was Julia. She was a doctor at Walter Reed Hospital. We talked about her specialty, cardiology, and what I was learning using gas-liquid chromatography.

That first ice-breaking experience led me to revise my schemata. New experience made clear to me that black people range widely in their occupations, interests, and activities. Julia also delighted me with her humor and helped me revise my undeclared fear that black people were mean, strict, and might even hurt me. Having discarded my schemata, I sought further opportunities to get to know people of color. I became aware of the ease I experienced in their presence, once I allowed myself to relax with them. Now, being with black friends reminds me of being at home in New Orleans, listening to jazz and laughing at other people's uptightness. New Orleans deserves its name "the city that care forgot." Black people seemed to invite me to take it easy. Of course, this perception constituted a new schemata, one that is continuously refined by ongoing life experiences.

Though my early fear of black people may seem incredibly out-of-date in today's climate of diversity and political correctness, it was not uncommon in 1963. I invite you to look back into your own life and ask yourself what schemata you have held and how they have changed over the years. Consider how prejudice in its many forms blocks us from creating cooperative communities—and a peaceful world. As cellular wisdom makes clear, prejudice is an aberration that defies the fundamental basis of our human sameness.

Unlearning Prejudice

My anecdotal experience suggests that one way of overcoming prejudice is seeking out experiences that challenge our implicit beliefs, while being willing to alter our schemata if evidence shows they are incorrect. Can such schemata-altering experience be effective with groups of people?

In one study, student volunteers participated in an experiment to examine whether education in prejudice and intergroup conflict can decrease negative orientation toward black people.[10] To begin, the students were given an inventory designed to measure implicit prejudice and anti-black stereotypes. The test required students to associate typically white or black first names with positive and negative attributes, such as *ambitious, industrious, successful, calm, trustworthy, ethical, lawful, lazy, shiftless, unemployed, hostile, dangerous, threatening,* and *violent.* A second part of the inventory measured associations between race and pleasant words, such as *sunshine, smile, angel, luck, rainbow, paradise,* and *fortune* and unpleasant words, such as *filth, death, devil, slime, cancer, hell,* and *poison.*

The coursework and discussions in the seminar were designed to foster respect for diversity. During the seminar, students were exposed to the dynamics of intergroup conflict. They engaged in discussions, sometimes heated ones, and kept a journal in which they documented evidence of bias, including their own. They were expected to increase their awareness of their degree of prejudice and of their level of motivation to counteract their biases.

After completing the seminar, students were again given an inventory. Both implicit prejudice—biases not likely to be disclosed to others—and explicit prejudice—stereotyping and other conscious attitudes—showed a marked decrease as a result of the seminar. Students who showed the most change in their attitudes reported being in strong agreement with statements about the course: "Made me realize African Americans still face a lot of prejudice and discrimination," "Opened my eyes to my own potential for biases and prejudice," and "Made me want to work harder at overcoming my own prejudices." They also agreed that the course "Allowed me to get to know/make friends with people outside my ethnic group."

As this study suggests, becoming aware of our biases appears to be important in reducing prejudice. Unless we are willing to face up to the prejudices we hold, we cannot choose to change our attitudes. A commitment to conscious introspection seems to be a prerequisite to finding ways to resolve conflict and enhance cooperation.

Overcoming Homophobia

The commonality of descent among the cells of the body does not always prevent problems. There are occasions when the body destroys itself because it doesn't recognize cellular elements as its own. Problems arising from this lack of recognition include autoimmune diseases, in which the immune system attacks the body's own cells, mistaking them for foreign invaders. These attacks cause a hypersensitivity reaction, as happens in allergies. An example of such self-attack is the disease multiple sclerosis. In this condition, the immune system attacks the myelin, the covering that insulates nerve fibers in the brain and spinal cord, causing inflammation.[11]

The social parallel to such internal attacks may be the homophobia and associated aggression that many gay people in the United States experience. Sixty-one percent of gay, lesbian, bisexual, transgender, and questioning students under the age of nineteen report verbal harassment, with close to 50 percent reporting experience it daily.[12] Statistics on the incidence of sexual harassment, physical harassment, and physical assault because of sexual orientation, as tallied in 2001 by the Sexuality Information and Education Council of the United States (SIECUS), are equally disturbing.[13]

Further evidence of the link between homophobic attitudes and aggression can be found in laboratory studies. In one experiment, participants were told that the study was aimed at understanding the effect on their reaction time in a competitive situation of watching sexually explicit material.[14] As part of the pre-screening tests for the study, participants' level of homophobia was determined by their responses to a series of tests, including the Homophobic Scale and the Aggression questionnaire. During the test, participants engaged in a contest against an opponent in which they were asked to press a button in response to a signal. The winner of each trial in the competition had the opportunity to administer an apparent electrical shock to the loser. The apparatus that delivered the shocks, called the aggression console, was a black metal box equipped with light-emitting diodes and electrical switches. The controls allowed participants to administer shock intensities rated from 1 to 5, five being the most intense. The students participating in the study believed that they were actually delivering electrical shocks to their opponent. In fact, the opponent was working as a confederate of the experimental team.

Seated in separate rooms, the subject and the opponent were each wired to the console. Both were shown a brief videotape of homosexuals engaged in sexual acts. The subject was also shown a videotape of his opponent and was able to listen to a conversation between his opponent and the researcher. In various trials, the opponent played the part of a homosexual man with stereotyped gay mannerisms, a gay man who described himself as being involved in a committed relationship for two years, or a heterosexual man involved in a committed relationship for two years. The researchers found that homophobic students were significantly more aggressive toward homosexual opponents than toward heterosexual ones, as evidenced when they delivered higher mean-shock intensities for a longer duration to opponents they perceived to be homosexual.

As experienced by people who have autoimmune diseases, aggression by one group against another within the body of society is a source of dysfunction and internal strife. Unable to recognize that sexual orientation is one of the many kinds of diversity that make up a pluralistic society, some people continue to believe that homosexuality is a mental disorder, an emotional problem, or evidence of a moral failing. However, the medical and psychological establishment firmly disagrees. The American Psychological Association now recognizes homosexuality and bisexuality as normal types of sexual orientation. Moreover, research data supports the claim that a preference for sexual partners of the same gender is more closely tied to genetic factors than to social conditioning.

A recent study focused on sexual orientation among twins.[15] In one part of the study, 951 pairs of twins were asked to describe their sexual orientation as heterosexual, homosexual, or bisexual. Data from identical twins—individuals who develop from a single fertilized ovum and therefore have identical DNA—were compared to that from fraternal twins—individuals who develop from two fertilized ova with different DNA—and with data from their non-twin siblings. Results showed that identical twins were more likely to have the same sexual orientation as compared to fraternal twins and non-twin siblings. The study suggests that genetic factors influence sexual orientation more than familial or social influences.

This data is consistent with the view that sexual orientation is not a choice. Homosexuals are a group of people whose DNA sequences, inherited from their parents, combined to form a unique DNA—as it true for each of us. For homosexuals, that particular combination of sequences supports a particular sexual orientation—just as inherited DNA sequences might support, say, musical talent or athletic ability. As in autoimmune diseases, homophobia is an inappropriate, hypersensitive reaction within the body of society, in which homosexuals are not recognized as belonging and are, therefore, attacked. Homosexuality is not a problem to be fixed; homophobia, like other examples of prejudice and conflict we have looked at, certainly is.

The depth of the suffering experienced as a result of homophobia among adolescents was documented by researchers who interviewed twenty gay men from a small town in South Yorkshire.[16] The interviewees described feelings of being different associated with their first understanding of being gay. For instance: "I remember going home at night and crying myself to sleep because I knew that I was different, and I was terrified" and "It made me an outcast." Being gay clearly challenges a young person's development of self-identity. The researchers recommend that schools foster an environment where it is easy for gay and lesbian teachers to be open about this aspect of their lives. Such positive role models can help gay and lesbian students in their process of identity formation.

MEASURING DIVERSITY AND PREJUDICE

Find a quiet place where you will not be disturbed and make yourself comfortable. Read each question slowly, allowing yourself sufficient time to understand it. Take your time reflecting on each question. Be responsive to any intuitive urgings you might experience. Remember that this is a completely private exercise. No one will see your responses, so you can be entirely truthful.

In answering the questions, think about your attitudes toward all groups — not only blacks and homosexuals, but also women, the disabled, and other ethnic and religious groups, such as Muslims, Palestinians, and Israelis.

Diversity Inventory

[] What minorities/majorities do you come into contact with on a regular basis?

[] In what context does this interaction occur?

[] What is the nature of the interaction?

[] Are you comfortable with this level and type of interaction?

[] If you wish to expand the diversity of your interactions, what steps might you take?

Prejudice Inventory

How would you describe your current attitude/feelings toward the following groups: blacks, whites, Latinos, Asians, Jews, Muslims, the disabled, homosexuals?

[] Has this attitude changed recently or over the years? In what ways? What events/experiences influenced this change?

[] Have you said or done anything to a minority member or group of which you are not proud? What were the circumstances?

[] How did these events affect your feelings/attitudes toward this minority or group?

[] What were the attitudes/feelings of your parents toward these groups?

[] What problems do you associate with these individuals or groups in the context of your interactions with them?

Fill in the blanks below with one of the minority groups listed above. Then consider whether you agree or disagree with the following statements. Repeat the inventory in regards to other minority groups.

If you give _____ an inch, they'll take a mile.

_____ have a chip on their shoulders, for no good reason.

If _____ really wanted a job, they would get one.

I had nothing to do with the fix that _____ find themselves in today.

In my view, _____ block the progress of others.

_____ think we're all the same.

_____ don't trust me, so I don't like or trust most of them.

Do you agree or disagree with the following statements:

Sex between two men or two women is wrong.

Marriages between gay and lesbian couples should be legalized.

Gays and lesbians should be able to adopt children.

Gays are more likely to molest children.

Gays and lesbians should be able to be open about their sexual orientation, even in the classroom.

Homosexuality is a conscious choice that can be voluntarily changed.

What did you learn about yourself as a result of this inventory? If you discover prejudice, ask yourself whether you wish to reduce it, and decide what action you might take. You may wish to join a group such as Teachers Against Prejudice, the Prejudice Institute, Citizens Against Homophobia, the National Organization for Women, or another organization devoted to reducing prejudice. Most such groups have websites, easily located using an Internet search engine.

Teaching Tolerance

What can be done at a community level to help people embrace diversity and overcome prejudice? One of the most hopeful initiatives is the Teaching Tolerance program, an initiative of the Southern Poverty Law Center. The program began in 1992, following a six-month study of the anti-bias resources available to schools and teachers. The study revealed that few appropriate materials were available to teachers, making the need for the program obvious and immediate. In the past ten years, the program has spread to schools across the United States. Teachers of grades K through 12 have access to program materials through *Teaching Tolerance* magazine, a publication that showcases innovative, replicable anti-bias activities used in schools across the country. Teachers and school administrators also have access to an award-winning series of curriculum packages developed by Teaching Tolerance, suitable for all grade levels.[17]

Park Day School in Oakland, California, puts the Teaching Tolerance philosophy into practice for students in grades K through 6. The school is committed to a diversified student body, supporting differences in ethnicity, race, religion, sexual orientation, and socioeconomic position. School administrators believe that learning in a diverse community is central to a child's ability to relate positively to others in our changing multicultural society. The school's curricula encourage children to "recognize the biases that exist in society and to develop and articulate their own values."

A teacher at Park Day School, Michelle McAfee, facilitates her students' appreciation of other cultures in her second grade classroom. Michelle spent a year in Montgomery, Alabama, on a fellowship in Teaching Tolerance research. Now back at Park, she structures her classroom around the model of a family, in which friendship, tolerance, and mutual support are offered to all members.[18] Her welcome wall displays an oil painting of a quilted bedspread. Each patch bears a student's name. A sign reading "Welcome to the Family" hangs below the painting. A wall hanging reflects the world cultures the students will learn about during the school year, from a Teaching Tolerance program called "Family Ties and Fabric Tales."

Michelle and her students discuss the fundamentals of supportive relationships necessary to a multicultural family, including cooperation, respect, empathy, and accountability. They work together to generate a list of classroom rights and responsibilities, which each student signs. Classroom problems are discussed in what Michelle calls family meetings, and the results are shared with other family members, including parents.

Similar support for diversity can be found in the school as whole. In 2002, Park sponsored a school-wide program centering on respect for differences in sexual orientation. Speakers were brought in from the San Francisco Bay area, including a lesbian animal caretaker and a lesbian minister. The San Francisco Gay Men's Chorus sang about self-pride. Tom Little, the director of Park Day School, explained the goal of program this way: "The idea is that if [the students] ever encounter homophobia, their association will be that lots and lots of wonderful people they have met have been hurt by that."

Classroom assignments reinforce the message of such programs. Students wrote essays on famous gay people like Michelangelo, interviewed a family member or family friend who is gay, and wrote Dear Abby–style letters to an imaginary gay child who experienced being teased. Classrooms were decorated with rainbow flags. One student spoke to first graders about his family, which consists of two moms. Imagine how much you love your mother, he told the first-graders—and then double that! Other programs at Park Day School center around other minority groups, such as a recent week devoted to the rights and problems of disabled people.

Clearly, Park Day School excels in creating opportunities for contact between people of various cultural heritages. In modeling a multicultural community, Park Day School prepares children for positive and harmonious contact with people who are different from themselves that they will encounter throughout life.

The Southern Poverty Law Center, founded by civil rights lawyer and community activist Morris Dees has sponsored other programs aimed at promoting diversity and combating prejudice. For instance, the Center has created a Citizen's Action Kit to help people initiate and participate in tolerance activities in their communities. Here are a few things we can do right now as suggested in *101 Tools for Tolerance:*[19]

- Attend a musical, theatrical, or dance performance by artists whose race differs from your own.

- Ask someone of another culture to teach you how to cook a traditional meal or invite someone from a different culture to dinner.

- Take a language course and practice conversing with someone in your community.

- When you hear racial slurs, let people know that you will leave the group should they speak like that again in your presence.

- Take the automated test to measure bias, available online at www.tolerance.org/hidden_bias/02.html.

- Volunteer in a Big Brother or Big Sister program.

- Create a multicultural calendar for home and school.

These tolerance initiatives reduce conflict and enhance creative cooperation as they promote the richness of diversity in our homes, schools, churches, places of business, and organizations. Like the community of the body, in which cells, organs, tissues, and systems work together cooperatively to keep us safe, healthy, and growing, the diverse groups in society can and must work together for the benefit of all. By expressing ourselves freely and accepting others, we have the opportunity to create peaceful, collaborative communities and, hopefully, a peaceful, collaborative world.

CHAPTER NOTES

1. Richard E. Michod and Denis Roze, "Cooperation and Conflict in the Evolution of Multicellularity," *Heredity* 86, no. 1 (2001): 1–7.

2. Alejandro Portes, "Social Capital: Its Origins and Applications in Modern Sociology," *Annual Review of Sociology* 24, no. 1 (1998): 1–24.

3. Muzafer Sherif et al., *Intergroup Conflict and Cooperation: The Robbers Cave Experiment* (Norman: University of Oklahoma Book Exchange, 1961).

4. Samuel L. Gaertner et al., "Reducing Intergroup Conflict: From Superordinate Goals to Decategorization, Recategorization, and Mutual Differentiation," *Group Dynamics: Theory, Research, & Practice* 4, no. 1 (1998): 98–114.

5. See note 4 above.

6. Henry T. Greely, "Human Genome Diversity: What About the Other Human Genome Project?" *Nature Reviews Genetics* 2, no. 3 (2001): 222–227.

7. Howard M. Cann, "Human Genome Diversity," *Comptes Rendus de l'Academie des Sciences—Series III—Sciences de la Vie* 321, no. 6 (1998): 443–446.

8. See note 6 above.

9. Paul DiMaggio, "Culture and Cognition," *Annual Review of Sociology* 23, no. 1 (1997): 263–287.

10. Laurie A. Rudman, Richard D. Ashmore, and Melvin L. Gary, "'Unlearning' Automatic Biases: The Malleability of Implicit Prejudice and Stereotypes," *Journal of Personality & Social Psychology* 81, no. 5 (2001): 856–868.

11. B. Mark Keegan and John H. Noseworthy, "Multiple Sclerosis," *Annual Review of Medicine* 53, no. 1 (2002): 285–302.

12. Anonymous, "Lesbian, Gay, Bisexual and Transgender Youth Issues," *SIECUS Report* 29, no. 4 (2001).

13. See note 10 above.

14. Jeffrey A. Bernat et al., "Homophobia and Physical Aggression Toward Homosexual and Heterosexual Individuals," *Journal of Abnormal Psychology* 110, no. 1 (2001): 179–187.

15. Kenneth S. Kendler et al., "Sexual Orientation in a U.S. National Sample of Twin and Nontwin Sibling Pairs," *American Journal of Psychiatry* 157, no. 11 (2000): 1843–1846.

16. Paul Flowers and Kati Buston, "'I Was Terrified of Being Different'": Exploring Gay Men's Accounts of Growing-Up in a Heterosexist Society," *Journal of Adolescence* 24, no. 1 (2001): 51–65.

17. www.teachingtolerance.org

18. Michelle McAfee, "Welcome to Park Day School," *Teaching Tolerance* 18, Fall (2000): 24–27.

19. See note 17 above.

CHAPTER 14

access genius

IN THE FIRST HALF OF THIS BOOK, we explored the cellular landscape within and worked to identify our values and articulate them. In the second half, we looked at the interactions between the systems of the body and worked on improving our human relationships and fostering more harmonious communities. This final chapter asks: What is the best we can be as individuals and as a community? And, what do we need to do to get there?

Most people would agree that the highest expression of human consciousness is genius—that remarkable combination of intellectual and personal qualities that allows a person to see beyond the obvious, to sense connections that escape narrower vision, and to express what they see in a way that ignites others and catalyzes change.

Who comes to mind when you say the word *genius?* Leonardo da Vinci? Thomas Jefferson? Marie Curie? Albert Einstein? All these are great geniuses, to be sure. But a book I first read many years ago, *The Man Who Tapped the Secrets of the Universe,* is about a genius who's considerably less well known, an example of genius, nonetheless.[1] Walter Russell was born in Boston in 1871. Because of financial reversals in his family, he left school when he was ten and took a job as a clerk in a dry goods store for $2.50 a week. Over the next several years, he taught himself to play the organ and, by age thirteen, he was working as a church organist. With the money he earned, he put himself through art school and became an illustrator for books and magazines. He also painted portraits, including one of the children of President Theodore Roosevelt. Although he never studied architecture, he designed the first Hotel Pierre in New York City, among other buildings. When he was fifty-six, he began to create sculptures of many famous figures, including Thomas Edison and Mark Twain. In his sixties, he won prizes for figure skating and continued to skate into his eighties.

Russell's philosophy of genius is simple: Mediocrity is self-inflected and genius is self-bestowed. If Russell is right, and anyone can be a genius, we might wonder: What role does the body/mind play in unfolding our genius? And, what qualities can help me to recognize my own genius and to access the genius of others?

As we see in what follows, the complex weave of interconnections that make up the "great intermediate net" of the nervous system is the essential physiological component of genius. This neural network allows for three qualities of mind clearly evident in the life of Walter Russell and every other great genius we might name: creativity, the ability to see beyond the obvious, shift perspectives, and explore ideas in new ways; interconnection, the ability to see relationships between ideas; and experience, the ability to learn from what has gone before, adapt to new circumstances, and extend what is known in new directions.

Walter Russell's ability to adapt to the changing circumstances of his life, to apply what he had learned about painting first to architecture and then to sculpture, and to exercise his creativity in so many fields—from music to art to figure skating—demonstrate how these three qualities of mind define a life of genius. Let's look now at what we can do to recognize and develop our own genius qualities.

The Great Intermediate Net

The qualities of genius depend on the ability to sense the environment and respond to it in unique ways. This ability developed slowly over our evolutionary history. Early in evolution, sensory cells, which detect things in the external environment, were directly connected to muscle cells. Under this arrangement, every stimulus led to the same response, the contraction of muscle fibers. Contracting muscle fibers allowed an organism to move toward or away from a stimulus—a response vital for its survival. As evolution progressed, neurons appeared, first simply as connectors between sensory cells and muscle cells. With continuous evolution, more and more complex organisms emerged with progressively larger numbers of neurons. As the neurons became more numerous, they coalesced into larger and larger interconnecting networks.

Eventually, these interconnecting networks came together to form the great intermediate net. Because of the net, each stimulus can elicit thousands of different responses. I might respond to a stimulus—say, the smell of freshly peeled figs—by remembering the row of fig trees on my Italian grandfather's farm. On another occasion, the same smell might remind me of the explosion of tastes as I bit into a fig. In fact, the same stimulus can elicit different responses from the same person at different times. Moreover, the smell of figs might elicit a completely different set of associations and responses from the person standing next to me. Because of the great intermediate net, the same stimulus can traverse any of thousands of pathways, activating one neuron and then another, and allowing me to bite into the fig or to see that it is overripe and throw it away.

This wonderfully responsive and flexible weave of interconnections is formed by all the neurons that are not directly involved in carrying incoming sensory messages to the spinal cord, or outgoing messages from the spinal cord to the muscles or glands. The net comprises the intermediary neurons that connect the neurons involved in direct sensory and motor function, and all neurons in the neocortex of the left and right hemispheres of the brain. The net links these newest structures of the brain to the older parts, such as the smell brain, the emotional brain, and the brain stem and spinal cord.

This great net is totally present to incoming stimuli, moment-by-moment, alert and ready to respond. It is engrossed in the moment and unaware of the future. Taken as a whole, its multiplicity of pathways transcends, by many orders of magnitude, the computing power of the direct link between neurons in the sensory and motor pathways. It would not be an exaggeration to say that the great intermediate net gifts us with our extraordinary capacities as human beings.

As impulses travel across neurons and between neurons along the network, I interpret my world. When I pick a fig, the familiar movements and smells activate neurons that store old memories and emotions. Each memory, in turn, activates responses from other neurons. I recall the texture of the skin on the outside of the fig. This memory reminds me of my grandfather teaching me how to tell a ripe fig from one that is overripe. A warm feeling spreads through my body as I remember how warm and nurturing my grandparents were.

The great intermediate net mediates our thoughts and feelings, memories and desires, beliefs and intentions. As the net's neurons pulse and communicate in multiple directions—up, down, this side, that side, across short axons, long axons, branches of axons—a synchrony of cellular wisdom emerges. The net is always ready. Pulsing with enormous potential, it awaits our direction.

To see the net in motion, watch a professional basketball player in the final series of the playoffs, after a season of honing his skills. Marvel at his coordination and speed as he dodges, swoops, dribbles, and scoops the ball into the basket. It's the net that integrates his sensory and motor neurons, keeping him in touch with his memories of the skill and experience of each player on each team. And it's the net that allows him to anticipate moves and strategies, while remaining aware of the time remaining on the shot clock and the position of each player on the court. Compare this multilevel awareness and repertoire of choices to simple sensory-motor reflexes, such as we experience when the doctor strikes the tendon just under the kneecap, and we involuntarily kick out our leg. Such responses are all that is available to the lower organisms from which we evolved. They would have found the genius necessary to play pro basketball quite beyond their reach!

Creativity

As we've said, creativity is the ability to see beyond the obvious, shift perspectives, and explore ideas in new ways. The key to thinking in this way seems to be staying in the present—seeing a problem as new, and exploring it and all its relationships without reference to the past. Staying in the moment allows new ideas and perspectives to emerge. Poet Mark Strand described his state of mind when he is writing: "Well, you're right in the work, you lose your sense of time, you're completely enraptured, you're completely caught up in what you're doing, and you're sort of swayed by the possibilities you see in this work. . . . [T]he idea is to be . . . so saturated with it that there's no future or past, it's just an extended present in which you're making meaning."[2] In fact, the process of staying in the present with a problem is short-circuited when we begin to evaluate our ideas.

One of the most important techniques of creativity is brainstorming, in which you allow the flow of creative associations to take you in many directions without judging which ideas are valuable and which are deadends. It's not difficult to see that the process of creative brainstorming is absolutely dependent on the great intermediate net. We've already noted that the intermediate net is totally present to incoming stimuli. Though it allows us access to past memories, its focus is reacting to what's happening right now, without an overlay of evaluation and self-judgment.

For instance, I often encourage my coaching clients to brainstorm about a problem they're trying to solve. Vanessa came to see me, quite upset. She was in the midst of a divorce. Her finances were limited, yet she needed to find an office to conduct her consulting business. I helped Vanessa recall times when she had been creative in solving problems. I assured her that a creative answer was waiting to be revealed to her; to find it, she simply needed to release any anxiety and worry and focus on the situation at hand. Then I asked Vanessa to describe her ideal office in detail. She began listing the features, such as the amount of space she needed, her desire for lots of light streaming in the windows, and a location close to the business center of town.

While such brainstorming is occurring, more and more neurons and neural pathways are being activated in the intermediate net. Consciousness acts like a spotlight, shining here and there, making connections, illuminating thoughts and memories, trying out possible solutions. As the process continues, more and more neurons are recruited, activating more of the great intermediate net.

Suddenly, Vanessa stopped speaking and looked at me, her eyes shining.

"What it is?" I asked her.

"My ex is moving to New York," she explained. "We are trying to sell our condo, but the real estate market isn't very good right now. I can ask my ex if we can wait to sell the condo for six months, until the market improves. During that time, I can use the condo as my office. It's perfect! I can't believe it—the solution was there all the time, but I didn't see it."

Have you ever had a similar experience? If so, did you notice an almost audible *click* when things fell into place? When Vanessa released her anxiety, she was able to see her situation from a new perspective, connect the condo she needed to sell with her need for an office, and solve her problem.

Just as the net makes creative problem solving possible for individuals, it can help corporations and communities to think creatively about their current situations. The same associative process that takes place in an individual involved in creative work can take place within a group of individuals who are working together to solve a problem. We might even say that business and community groups can function as an "extended intermediate net" linked by a complex network of social and corporate interconnections.

For example, Pitney Bowes Credit Corporation (PBCC), a division of Pitney Bowes, a manufacturer of office supplies, recently reinvented itself by adopting a corporate model that fosters just this kind of associative, creative thinking. Matthew Kissner, PBCC's president and CEO, calls his division an idea factory.[3] His first week on the job, he handed out buttons with the words *That's the way we've always done it* in a red circle, crossed out with a slash. Instead of adhering to PBCC's traditional business—financing the sale and leasing of equipment—Kissner refocused the division on creating customer services, such as establishing a credit card geared to small businesses. Because the majority of PBCC's customers were small businesses, the card proved to be a popular service and PBCC's revenues shot up.

To encourage teamwork and creativity, Kissner moved his division into a building designed to encouraged brainstorming and collaboration. White boards outlining the goals for the quarter appeared in every department. Traffic signs included words that reminded workers of strategies for success, such as *Grow the customer base* and *Capitalize on the power of people*. Instead of an annual retreat open only to high-ranking employees, Kissner made the gathering an attractive perk for employees whose performance during the year was exceptional. In Kissner's words, he wanted "people to bump into each other, talk about what they're doing, and exchange information that they wouldn't otherwise exchange."[4] Employees were encouraged to be information gatherers. One area of the building, called the Cranial Kitchen, featured booths where employees could watch training videos or surf the net.

The results of this emphasis on creativity were astonishing. In 1998, the reinvented division brought in 36 percent of the profits for Pitney Bowes, although it accounted for only 2 percent of the parent corporation's workforce. One of its new services, a revolving line of credit that helped small businesses finance their postage costs, attracted four hundred thousand customers within the first nine months! The creativity apparent at PBCC demonstrates, on a corporate scale, the marvels that are available to all of us when we make the effort to access the innate potential of our creative genius.

A similar kind of cooperative effort can help communities solve their problems. In 1997, a massive flood of the Red River carved a five-mile-wide path through Grand Forks, North Dakota, and East Grand Forks, Minnesota.[5] Every business was flooded, and only a few houses were spared. In the aftermath of the flood, many people wanted to turn their backs on the river and rebuild their lives elsewhere. But others rallied behind the idea that the river could be tamed and turned into an asset for the communities. These folks consulted a Massachusetts company for answers. With the company's help, they built a one-thousand-foot wall protecting the downtown areas and supplementing two dikes that protected the rest of the city. As inspiration for the reconstructed downtown, the residents looked to the past—to the Roaring Twenties, when East Grand Forks was a local entertainment center. They lobbied the state legislature for liquor licenses and mapped plans for restaurants facing the river. A historic saloon became the centerpiece, rebuilt as the art deco landmark. Now there are offices and boutiques in the flood-protected downtowns, and the cities are once again thriving centers for business, shopping, and entertainment. Today the people of Grand Forks and East Grand Forks say they're living in the luckiest cities in the world.

Possibilities for personal and communal growth abound when we have the courage to look beyond the ordinary to find them.

ACCESSING YOUR CREATIVITY

Find a place and time that is conducive to exploring your inner landscape. Surround yourself with things that might help you to remember your genius, such as an award you were given, a photo of a special event, a project, or piece of art you created. First, consider the following question:

[] When was the last time you devised a creative solution to a problem or a new way of doing something that surprised others? What do you remember about this occasion? How did this event make you feel?

[] Do you ever have inspirations but not act on them? When was the last time that happened? What stopped you from acting on the inspiration?

[] The first step in accessing your creativity is giving yourself permission to do so. Do that now. Remind yourself of how you feel when you follow through on a creative idea. Choose several activities from the following list to help you anchor your commitment to creativity:

[] Read a book on creative thinking, such as *Cracking Creativity: The Secrets of a Creative Genius,* by Michael Michalko;[6] *Creativity: Flow and the Psychology of Discovery and Invention,* by Mihaly Csikszent-mihalyi;[7] *Discover Your Genius: How to Think Like History's Ten Most Revolutionary Minds,* by Michael J. Gelb;[8] or *Think Like A Genius: The Ultimate User's Manual for your Brain,* by Todd Siler.[9]

[] Keep a creativity scrapbook in which you place anything that catches your attention—photos, quotations, fabric, your own drawings or poems, an ad from a magazine. Take the scrapbook out from time to time and brainstorm about different ways to use what you've collected to create something new: a card for a special occasion, a gift for a friend, an illustration to accompany a poem, or some other project.

[] Whenever possible, do something you need to do in a new way. Drive to work using a different route. Go to a play instead of a movie. Write or eat with your non-dominant hand. Cook a new kind of ethnic food.

[] Try out your commitment to creativity by exploring an issue or problem in your life that is unresolved.

[] Give yourself three weeks or more to explore the issue or problem.

[] During the first week, spend ten minutes each day brainstorming. Write down whatever ideas come without evaluating them.

[] At the end of the first week, engage in a physical activity like dancing, or a centering activity like yoga or meditation. When your mind is clear and you are in an expansive state, peruse what you have written. Use a highlighter to mark any words or phrases that draw your attention.

[] Write these words or phrases on a separate piece of paper. Draw lines between words that seem to go together or group them under larger headings.

[] Spend the second week continuing to brainstorm about the idea.

[] At the end of the second week, make a mind map. Draw a circle in the center of a piece of paper. Write the issue or problem in the circle.

Now take any big ideas or headings that have emerged from your previous brainstorming sessions and place each in a circle on the same page.

[] Connect these circles to your main circle and to each other, as seems appropriate. Allow other ideas to emerge. Write them in their own circles connected to any of the circles on your page.

[] Keep this mind map in a place where you can see it. As new ideas emerge, add them to the map.

[] At the end of the third week, sit down with your mind map and a fresh piece of paper. After studying the mind map, write three or more possible solutions to your problem on the fresh piece of paper. Choose one and list the steps you might take to implement it.

[] Make the commitment to put the first step of your plan into action, and then do it!

[] Consider whether there is a problem in your family, neighborhood, organization, or workplace that could benefit from creative thinking.

[] Bring together a group of people who are affected by this problem to brainstorm a solution.

[] Use an easel or whiteboard to record everyone's ideas. Come to an understanding that all ideas will be recorded without evaluation or judgment.

[] At a second meeting, or after a break for refreshments, set the brainstorming list aside and engage in a session of mind mapping, following the model given above.

[] Finally, list the possible solutions to the problem and discuss them, with the goal of reaching consensus on a plan of action.

Interconnection

In addition to the ability to see beyond the obvious, a genius is also gifted at seeing relationships between ideas and concepts—and at forging connections between people. According to psychologist Mary-Elaine Jacobsen, many of us are everyday geniuses—people who are especially gifted at perceiving relationships and making such connections.[10] Jacobsen explains that everyday geniuses know that there is an interconnectedness in all things. What is interesting or puzzling to them now almost always becomes relevant later. For instance, she tells the story of Samuel Morse, who struggled to find a way to make his Morse code signal powerful enough to transmit over long distances. One day, while watching the changing of horses at a Pony Express relay station, he came up with the idea of periodically boosting the power to his traveling signal at electronic relay stations—just the solution he needed to complete the invention of the telegraph.

Geniuses take advantage of chance and look for multiple ways of viewing things. The urge to find connections drives them to pursue ambiguities relentlessly. Because they think in terms of words, numbers, spatial relationships, and visual images, geniuses generate more ideas than thinkers who use a single approach. Intrigued by complexity and double meaning, geniuses look, research, look some more, pull up stored material from experience, doodle, draw, chart, and diagram to expose relationships. Moreover, everyday geniuses work at what they love and allow themselves to be guided by their own fascinations. Jacobsen also tells the story of Jerry Yang and David Filo. When the two were at Stanford in the early 1990s, they loved to play on the Web. However, they were hampered by not being able to call up Web addresses readily. So, in 1994, they created a listing and shared their "David and Jerry's Guide to the World Wide Web" with others. At that moment, Yahoo! was born.

Many of us can recall times, especially in childhood, when making connections and thinking creatively was easy and fun. However, a budding genius can easily be discouraged by parents and teachers. Jacobsen recounts that as a child, she was often given conflicting messages. Sometimes, she was told she was special; at other times, she was admonished for being too curious and too inquisitive. By age five, she says, she had learned that it was important to hold back and to be on guard. Experience had taught her that when she followed the rules, she was safe. For instance, no one knew she wrote poems for the fun of it. When she took a chance and dared to express herself, the cost was high. Her fifth-grade teacher shredded a poem she had written in front of the class, saying, "No one your age could have written such a thing. Shame on you." Our creative genius at making connections is also stifled by messages such as "Can't you just stick to one thing?" "Where do you get those wild ideas?" and, of course, "Who do you think you are?" Maybe you've heard similar things or have said them to your children. The ones I heard repeatedly while growing up were, "Be practical" and "Why do you have to do everything the hard way?"

In midlife, Jacobsen recounts, she realized that she had lost something vital. Remembering an early professional interest in gifted adults, she shared this interest with a professor, who replied, "Of course you're interested in gifted people; you are one." These words, Jacobsen writes, were a "bombshell that blasted me into a personal cycle of newfound energy, possibility and confidence that has not slackened to this day." Looking back, each of us can probably point to some moment when our innate curiosity about the world and the urge to discover for ourselves how things fit together was squelched. Many of us have adopted a safe way of living, keeping our inquisitiveness and our wild ideas under wraps. Wouldn't it be wonderful if we could each be blasted by such a bombshell revelation and rediscover the irrepressible genius that is our birthright?

Creative thinkers direct their attention to a broad range of stimuli, storing seemingly irrelevant information and recalling it later when it becomes useful for solving a problem. In one study, psychologists gave college students ten minutes to memorize a typed list of twenty-five words.[11] As they studied the list, one of the test-givers read another list of words to them. The students were instructed to ignore the words being read aloud. When the study period was over, the students were given a series of thirty anagrams—words or phrases formed from others by rearranging the letters. For example, *angel* is an anagram of *glean*. The students were given ten seconds to solve each anagram. What the students did not know was that some of the solutions to the anagrams were included among the list of words they had been asked to memorize; others were among the words read aloud to them.

After solving the anagrams, the students were asked to recall as many of the words as they could from the list they had been asked to memorize, and any of the words they could remember from the list that had been read aloud to them. Students who recalled more of the words that had been read aloud solved more of the anagrams—even though they had been instructed to ignore those words. The most creative thinkers in the group stored a memory of the words and used them later to solve the anagrams. The authors concluded that people who focus their attention more diffusely tend to make more unusual connections. This ability seems to be an essential element of creative thinking.

Coupled with the ability to make connections between ideas is the talent for recognizing and fostering relationships between people. William E.

Strickland Jr., founder of Pittsburgh's Manchester Craftsmen's Guild and other agencies dedicated to social change, is a genius at this kind of inter-connection. Strickland founded the Guild in 1968, when he was still in college, as a program to teach pottery to children in Manchester, the decaying Pittsburgh neighborhood where he grew up.[12] He believed that crafts, such as pottery, can teach children creativity and the skills for business success, because he learned these skills under the mentorship of a ceramics teacher at his Manchester high school. Strickland describes seeing this teacher mold a mound of clay into a useful and beautiful vessel for the first time as a "clairvoyant experience"—a vision of how the world ought to be. He was determined to provide this lesson in practical creativity to others.

The aim of the Craftsmen's Guild is not to make children into artists, but rather to teach them, person to person, that they are innately creative, that they can set goals and achieve them. Strickland believes that "artists are by nature entrepreneurs. . . . They have the ability to visualize something that doesn't exist, to look at a canvas and see a painting. Entrepreneurs do that." To that end, he offers children programs in ceramics, photography, computer imaging, and drawing, who clearly thrive in the creative and nurturing environment he has created. In 2000, Strickland reported that 80 percent of the at-risk children who participate in the Guild's programs go on to college.

Another of Strickland's enterprises, the Bidwell Training Center, extends this visionary approach to adult education, providing superb professional training free of charge to people on welfare, giving them the chance for a second start. Among the training courses offered at Bidwell is a program in culinary skills, including nutrition and food management. One of the students in this program, Janine Johnson, a single mother of four on welfare, is studying to be a chef. Part Cherokee, her dream is to open a restaurant called Janine's Four Winds Reservation.

Strickland's latest venture is a program to teach effective, profit-generating management skills to executives of nonprofit organizations, such as arts centers, museums, and symphony orchestras. The program is called the Denali Initiative, named after the highest peak in the United States because it aims to take people to a higher summit. Participants in the program spend three years learning to manage a nonprofit in a way that generates sustainable revenues while having a positive social impact.

Strickland's genius for connecting people with each other and with their dreams for the future has been recognized as a model for visionary social change. Harvard Business School, where he has presented his ideas, calls him a social entrepreneur. In 1996, Strickland was awarded a MacArthur Foundation Genius Award.

Like Strickland, Brenda Krause Eheart, professor of sociology at the University of Illinois, sees connections where others do not. Eheart's dream was to create a community that would support families who were willing to adopt "unadoptable" children—older children, children scarred by abuse and neglect, minority children, sibling groups, and children with behavioral and emotional problems. In 1994, Eheart created a nonprofit corporation that purchased part of the closed Chanute Air Force Base in Rantoul, Illinois, to establish a community called Hope Meadows.[13]

Hope Meadows is a twenty-two-acre housing subdivision, with single-family homes, apartments, and administrative and community buildings. The five-block neighborhood, with its tree-lined streets, resembles a community from the 1950s. It aims to be an "idyllic semi-rural, working-class environment, where kids walk to public schools, ride bicycles safely, and run freely between and around the houses from one large open greenspace to the next." As of 2001, thirty-one foster children were living at Hope Meadows, ranging in age from days old to thirteen years. Many of the children, about two-thirds, had spent over half their lives in foster care.

Parents at Hope Meadows who adopt four children from the foster care system receive a six- to seven-bedroom home, rent-free. In addition, they receive about nineteen thousand dollars a year in stipends, enabling one parent to remain in the home as a full-time mom or dad. Support for these adopting families includes assessments and evaluations of the children, child welfare services, weekly parent training, and on-site counseling for both children and their families.

Connecting parents with children who need them is one aspect of Eheart's genius. But she and her associates also recognized that a supportive intergenerational community is essential for special-needs children. So, she rents apartments at Hope Meadows to fifty-nine older adults. These seniors live in air-conditioned apartments for which they pay no more than $350 a month in rent. They contribute to the community by working a combined total of fourteen hundred volunteer hours per month. Often they can be found in the Intergenerational Center, with its children's library, computer room, rooms for individual tutoring, kitchen, and large open meeting space. The volunteer grandparents help the children of Hope Meadows with their homework, play cards or board games with them, help them use the computers, and coach and referee their soccer or basketball games. The seniors also help care for the younger children, guard school crossings, and supervise playgrounds. Most important, they become part of the children's lives, listening to them and sharing their wisdom and insight.

Hope Meadows is interracial, intergenerational, accepting of individual differences, safe, and supportive for all its residents. The interconnections it fosters—between children and parents, younger and older residents, and families and supportive professionals—is a model for the communities where most of us might wish to live, in which every person contributes for the benefit of all.

The success of this project has been recognized with numerous honors. In 1997, Brenda Krause Eheart was one of *Ms.* magazine's "Women of the Year." In 2000, she was granted Oprah Winfrey's "Use Your Life" award, and in 2001, she received *People* magazine's "Heroes Among Us" award. The senior volunteers at Hope Meadows have also been recognized. In 1999, they won the Illinois Governor's Home Town Award for Volunteerism.

Experience

A third aspect of genius is the willingness to learn from what has gone before and extend what is known in new directions. Because the great intermediate net gives us access to what we have already learned while keeping us rooted firmly in the present, we have the opportunity to use our experience to better our own lives and the life of the communities in which we live. Creating a vehicle for channeling the experience of individuals back into the community has been a passion for several remarkable people.

Marc Freedman and John W. Gardner used their experience and their network of connections to build Experience Corps, a network of volunteers who use their life experiences to improve the education system, social services, and neighborhood living.[14] Freeman founded Public/Private Ventures and Civic Ventures, two nonprofit organizations dedicated to developing innovative strategies that help disadvantaged children. Gardner, founder of Common Cause and the former Secretary of Health, Education, and Welfare under President Lyndon Johnson, was one of America's greatest social thinkers and entrepreneurs. He died in February 2002, but the service organizations he co-founded continue to thrive.

Experience Corps operates in fifteen cities throughout the United States and has recently expanded to the United Kingdom. The program provides adults fifty-five years of age and older with the opportunity to help children improve academically and to guide them as they develop. Volunteers (called members) participating in the program typically spend fifteen to twenty hours a week in elementary schools, particularly in the inner cities, tutoring and mentoring children; supporting teachers; encouraging and supporting parent involvement; and reviving and staffing school libraries. Volunteers also develop enrichment programs, including before- and after-school programs devoted to music, sports, and dance. One of the goals of Civic Ventures, a social service organization that serves as the central office for Experience Corps, is to extend the seniors volunteer program to venues other than schools, such as YMCAs, Boys and Girls Clubs, and other community youth organizations.

In addition to the clear gains for the children served by Experience Corps volunteers, people who work for the Corps also report tremendous personal benefits. Their inspiring stories are told on the Corps' website, www.experiencecorp.org. Ida May Hughes, for instance, has been working in the Kansas City Experience Corps program since 1998. She mentors teenagers at the Linwood YMCA, talking with them about life and teen issues and trying to improve their lives. She has become, she says, "both substitute mother and grandmother to many of the young people."

Hughes was a caregiver for her husband, until he died in 1998 of Lou Gehrig's disease. Her work for the Corps helped in her own healing process. Although she is in her seventies, Hughes says she feels as though she is still in her twenties, thanks to "keeping busy" and serving others. She brings to her work thirty-five years of experience as a scout leader, and the wisdom gained from being the mother of four children, the grandmother of twelve, and the great-grandmother of ten. Evidence of the positive personal effect of her work for the Corps is that Hughes plans to return soon to Penn Valley Community College to finish her bachelor's degree in early childhood education.

Also working for the Corps is Ed Blystone, a retired truck driver and former Teamster from Chicago. Blystone moved from Chicago to Portland, Oregon, where he spent four years homeless and living on the docks. After nearly dying of pneumonia, he decided to dedicate the rest of his life to serving others. His description of his work for the Corps testifies to the genius of experience: "I am a different person [than I was thirty or forty years ago]. That comes with age, I think, and wisdom. When you get older, you get wiser. You look back at all of those young years and think, 'Why wasn't

I this smart then?' . . . I really believe that I've made a difference in a lot of little kids' lives. I give a lot of those little kids love that they don't get. . . . I try to be a role model for them."

Though he describes himself as "salty and hard," sometimes the affection shown by the children makes him "want to cry," such as the time he "had one of 'em come up and throw her little arms around me and say, 'Grandpa Ed, I love you.'"

A similar initiative to Experience Corps is Volunteers in Medicine, founded by Dr. Jack McConnell, a physician turned pharmaceutical company executive. This service organization harvests the experience of retired doctors, nurses, dentists, and other health-care professionals to provide free health care for the uninsured working poor and their families. McConnell tells the story of having picked up a man who was standing in the pelting rain with no umbrella and no raincoat. As they drove, McConnell asked the man where he was going. He learned that the man had just been laid off a construction job and was looking for work to support his wife and two children. McConnell was not able to resist asking the man if his family had access to medical care. No, the man replied. The only time his family had had decent health care was when he had been in the Army. That conversation inspired McConnell to do something constructive to help similar families. The first Volunteers in Medicine clinic was opened in Hilton Head, South Carolina, in 1992. Today, there are clinics in ten states.

McConnell has received nearly five hundred requests from communities in the United States asking for guidance in creating similar clinics. To meet this need, he created the Volunteers in Medicine Institute, which sponsors nationally recognized speakers, provides on site and phone consultations, and creates and distributes resource materials, including a start-up guide, a clinic support manual, and a video on creating a culture of caring.

Experience Corps and Volunteers in Medicine provide powerful testimony to the ways the genius of creativity, interconnection, and experience can help us build more humane and caring communities.

SHARING OUR GENIUS TO CREATE CARING COMMUNITIES

Find a comfortable spot that encourages reflection. Allow thoughts about the stories you have read to linger in your mind and heart. Now bring to mind the communities of which you are a member. Go inside and ask yourself:

[] Have I ever had the dream of making a contribution in service to others in my community?

[] Did I follow through on this dream? Why or why not?

[] If I did not follow through, what thoughts or feelings stopped me?

[] What current needs do I see for service to others in my community?

[] What creative ideas, networking possibilities, and experience might help me fulfill these needs?

[] Who would I need to help me use these ideas or skills?

[] What steps am I willing to take to put this dream into action?

Take your time exploring these questions. The more specifically you can define the need you wish to address, the easier it will be to take action. If you wish, use the brainstorming and mind-mapping techniques you learned earlier in the chapter to clarify what you hope to accomplish. At the end of this process, jot down a list of three specific goals, making sure that they are realistic given your skills and experience. Be sure to link these goals with a clear time line, so that you can measure your progress toward achieving them. Remind yourself that by serving others, you are benefiting yourself as well. Like the inspiring people you have read about, you too can be a genius and encourage and support the genius of others!

]0000[

WHEN YOU ARE ASKED, "Who do you think you are?" recall the resource that resides within you—the great intermediate net. It is ever-present, awaiting your call! Authentic desire, anchored in choices aligned with the truth of who you are and who you can be, activates the net. You do not need to leap tall buildings in a single bound. A day-to-day focus on truth, a clear eye and heart, is all that's needed to unfold your genius.

Relationships thrive when you deal authentically with others, showing them caring concern. The great intermediate net is present in all people. Help others to know who they are, and everyone thrives. Communities flourish when you have the courage to reflect their truth back to them and to speak up when they no longer act in accordance with that truth. The universal energy that we have named cellular wisdom can be your inner teacher. All you need do to access its genius is to look inside.

CHAPTER NOTES

1. Glenn Clark, *The Man Who Tapped the Secrets of the Universe* (Waynesboro: The University of Science and Philosophy, 1975).

2. Mihaly Csikszentmihalyi, *Creativity* (New York: HarperPerrennial, 1996).

3. Scott Kirsner, "Designed for Innovation," *Fast Company* 19 (1998): 54–56.

4. See note 3 above.

5. Linda Tischler, "Grand Forks and East Grant Forks: After the Flood (Literally)," *Fast Company* 60 (2002): 84–86.

6. Michael Michalko, *Cracking Creativity* (Berkeley: Ten Speed Press, 2001).

7. See note 2 above.

8. Michael J. Gelb, *Discover Your Genius: How to Think Like History's Ten Most Revolutionary Minds* (New York: Harper Collins, 2002).

9. Todd Siler, *Think Like a Genius: The Ultimate User's Manual for Your Brain* (New York: Bantam Books, 1996).

10. Mary-Elaine Jacobsen, *The Gifted Adult: A Revolutionary Guide for Liberating Everyday Genius* (New York: The Ballantine Publishing Group, 1999).

11. Pamela I. Ansburg and Katherine Hill, "Creative and Analytic Thinkers Differ in Their Use of Attentional Resources," *Personality and Individual Differences* (forthcoming).

12. Sara Terry, "Genius at Work," *Fast Company* 17 (1998): 171–181.

13. David Hopping, Martha Bauman Power, and Brenda Krause Eheart, "Hope Meadows: In the Service of an Ideal," *Children and Youth Services Review* 23, no. 9-10 (2001): 683–690; Brenda Krause Eheart and David Hopping, "Generations of Hope," *Children and Youth Services Review* 23, no. 9–10 (2001): 675–682.

14. Marc Freedman, *Prime Time* (New York: Public Affairs, 1999).

CONCLUSION

Congratulations! In reading *Cellular Wisdom,* you've begun exploring, peering into your body, and decoding its messages. As you've engaged in doing the chapter exercises and subsequently implementing them, you've begun an ongoing process of creating your unique and authentic expression—step-by-step, moment-by-moment. As we explored how neurons communicate in chapter 2, you discovered that synaptic transmission requires only milliseconds to occur. In contrast, all of us are familiar with the decades required for the growth of bones—from birth to maturity. Similarly, you may find that applying the inner teachings to your life can require as varied a time span, from seconds to decades.

Consider for example, Bertha, who first engaged in coaching with me to advance in her profession. A successful marketing expert, Bertha's personal life was much less successful. Though she had been living with Mark for ten years, she felt uncomfortable in the relationship. In probing her relationship, I explained the principle of cellular wisdom—of living from the inside out from her unique blueprint. Almost instantaneously, Bertha recognized the source of her discomfort. She had tried to make the relationship fit. As she reflected on her values and her essential goals, the perspective from which she viewed the relationship changed instantly. She recognized something that she had known for years, but not allowed herself to see: the disparity in values between herself and her partner. Like the milliseconds required for neurotransmission, clarity blazed! Bertha recognized her relationship as one that she needed to terminate that moment. She came to the decision to break off her ten-year relationship by the time she left my office. The decision made, relief flooded through Bertha. The ever-present knot in her gut loosened. The persistent theme of "Yes, it will work—it has to" reverberating through her mind was silenced. She felt her spirit soar. Relief remained present even as she went about discussing her decision with her partner, finding a new apartment, moving, and so on.

Within a few weeks, a different aspect of Bertha's life became the focus of her attention. This time, however, the process of implementing the teaching of her unique blueprint required a considerably longer time to unfold. Years before, Bertha had wanted to go to law school. However, she had put away her dream. Her parents thought law school would be too difficult for her; instead she had taken a job in an advertising agency. Spurred by her move, the idea of law school re-emerged. The comments and concerns of her friends and family, such as Why would anyone want to go back to school at this point in life? now served as promptings for her to connect with her inner truth. Almost a year later, celebrating her fortieth birthday, Bertha announced that she would be going to law school part-time in the evenings and on weekends. She interviewed several graduates of the law school who had followed this path and learned about the obstacles they faced. Cellular teachings generated a new perspective through which Bertha could view the challenge of law school. Over a period of days, weeks, and months she unfolded a practical plan to achieve her goal using her inner guide. Her friends commented that she radiated vitality.

You may experience a different scenario than Bertha's, as you listen to your inner teacher. Life may stop, come to a halt. The structures of your life, marriage, family, relationships, and work, may collapse. You cannot charge ahead. You cannot see where to go. Chaos reigns. Take a deep breath. Reflect on cellular wisdom. Let yourself realize the process must unfold naturally, organically. It may sometimes proceed as slowly as the healing of a broken bone. Access your own inner wisdom. Be reassured. The process will unfold, if you allow it. The body announces in all of its cells and systems— change is inherent to life. The light within you will shine on a path. Reflect on the inner teachings here presented. Let them stew in the substance of your being. Incubate them. Their meaning will emerge. When this understanding has matured, you will experience its clarity and witness many synchronicities. The body's teachings will light your path. Follow them to an exuberant life of true self-expression.

How can you best extract the teachings to make a difference in your life, regardless of the scenarios that present themselves to you? My clients have shown me a variety of approaches that are effective. Trust your inner promptings and follow the approaches that you find most enticing. These will probably contribute most to your development. You can reread those parts of *Cellular Wisdom* that were most meaningful to you. Another approach is to simply open the book at random and reflect on the teachings contained on the open pages. Perhaps you'd like to redo the exercises or explore them in greater depth than you did originally. Journal your reflections. Catch your insights before they sink into the sea of obscurity. Do the meditations on www.cellular-wisdom.com while viewing the magnificence of cells, seen in their glory in real images. Resonate with their truth. I've found that discussing a novel perspective, such as that of cellular wisdom, with others in a group setting can reveal hidden gems. Gather a group of like-minded individuals. Inquire and explore cellular wisdom. Do the exercises together. Discuss what emerges. If this method works best for you, don't be alarmed; sometimes the energy of the group allows you to recognize your answers and resonate with them. You are the director of your life. And you are never alone. Realize you carry within you your own wisdom, the wisdom of your cells.

BIBLIOGRAPHY

Chaleff, Ira. 1998. *The Courageous Follower: Standing Up to and for Our Leaders*. San Francisco: Berrett-Koehler Publishers.

Csikszentmihalyi, Mihaly. 1996. *Creativity: Flow and the Psychology of Discovery and Invention*. New York: HarperCollins.

de Chardin, Pierre Teilhard. 1999. *The Human Phenomenon*. Portland, Oregon: Sussex Academic Press.

Freedman, Marc. 1999. *Prime Time: How Baby Boomers Will Revolutionalize Retirement and Transform America*. New York: PublicAffairs.

Gelb, Michael J. 2002. *Discover Your Genius: How to Think Like History's Ten Most Revolutionary Minds*. New York: Harper Collins.

Gladwell, Malcolm. 2000. *The Tipping Point: How Little Things Can Make a Big Difference*. Boston: Little, Brown and Company.

Jacobsen, Mary-Elaine. 1999. *The Gifted Adult: A Revolutionary Guide for Liberating Everyday Genius*. New York: The Ballantine Publishing Group.

Kyle, David T. 1998. *The Four Powers of Leadership: Presence, Intention, Wisdom, Compassion*. Deerfield Beach, Florida: Health Communications, Inc.

Maslow, Abraham H. 1962. *Toward a Psychology of Being*. New York: Van Nostrand Reinhold.

Michalko, Michael. 2001. *Cracking Creativity: The Secrets of a Creative Genius*. Berkeley: Ten Speed Press.

Michod, Richard E. 1999. *Darwinian Dynamics*. Princeton, N.J.: Princeton University Press.

Siler, Todd. 1996. *Think Like a Genius: The Ultimate User's Manual for Your Brain*. New York: Bantam Books.

Vaillant, George E. 2002. *Aging Well: Surprising Guideposts from the Landmark Harvard Study of Adult Development*. Boston: Little, Brown and Company.

Vertosick Jr., Frank T. 2002. *The Genius Within: Discovering the Intelligence of Every Living Thing*. New York: Harcourt, Inc.

INDEX